REDUCTIONISM AND SYSTEMS THEORY IN THE LIFE SCIENCES

THEORY AND DECISION LIBRARY

General Editors: W. Leinfellner and G. Eberlein

> Series A: Philosophy and Methodology of the Social Sciences
> Editors: W. Leinfellner (Technical University of Vienna)
> G. Eberlein (Technical University of Munich)
>
> Series B: Mathematical and Statistical Methods
> Editor: H. Skala (University of Paderborn)
>
> Series C: Game Theory, Mathematical Programming and
> Operations Research
> Editor: S. H. Tijs (University of Nijmegen)
>
> Series D: System Theory, Knowledge Engineering and Problem
> Solving
> Editor: W. Janko (University of Economics, Vienna)

SERIES A: PHILOSOPHY AND METHODOLOGY OF THE SOCIAL SCIENCES

Volume 10

Editors: W. Leinfellner (Technical University of Vienna)
G. Eberlein (Technical University of Munich)

Editorial Board

Scope

This series deals with the foundations, the general methodology and the criteria, goals and purpose of the social sciences. The emphasis in the new Series A will be on well-argued, thoroughly analytical rather than advanced mathematical treatments. In this context, particular attention will be paid to game and decision theory and general philosophical topics from mathematics, psychology and economics, such as game theory, voting and welfare theory, with applications to political science, sociology, law and ethics.

For a list of titles published in this series, see final page.

REDUCTIONISM
AND SYSTEMS THEORY
IN THE LIFE SCIENCES

Some Problems and Perspectives

edited by

PAUL HOYNINGEN-HUENE
Technical University of Zürich

and

FRANZ M. WUKETITS
University of Vienna

KLUWER ACADEMIC PUBLISHERS
DORDRECHT / BOSTON / LONDON

Library of Congress Cataloging in Publication Data

Reductionism and systems theory in the life sciences : some problems
  and perspectives / edited by Paul Hoyningen-Huene and Franz M.
  Wuketits.
        p.   cm. -- (Theory and decision library. Series A, Philosophy
  and methodology of the social sciences)
    Includes index.
    ISBN-13:978-94-010-6941-0   e-ISBN-13:978-94-009-1003-4
    DOI: 10.1007/978-94-009-1003-4

    1. Biology--Philosophy. 2. Reductionism.   I. Hoyningen-Huene,
  Paul, 1946- II. Wuketits, Franz M. III. Series.
  QH331.R39 1989
  574'.01--dc20   89-15308

ISBN-13:978-94-010-6941-0

Published by Kluwer Academic Publishers,
P.O. Box 17, 3300 AA Dordrecht, The Netherlands.

Kluwer Academic Publishers incorporates
the publishing programmes of
D. Reidel, Martinus Nijhoff, Dr W. Junk and MTP Press.

Sold and distributed in the U.S.A. and Canada
by Kluwer Academic Publishers,
101 Philip Drive, Norwell, MA 02061, U.S.A.

In all other countries, sold and distributed
by Kluwer Academic Publishers Group,
P.O. Box 322, 3300 AH Dordrecht, The Netherlands.

TABLE OF CONTENTS

PREFACE

The present volume aims at giving a discussion of the problems of reductionism in contemporary life sciences. It contains six papers which deals with reduction/reductionism in different fields of biological research. Also, the holistic perspective, i.e. the systems view, is discussed in some of the papers. The message of this discussion is that – whereas reductionism is indeed an important strategy – the systems approach is needed. It is argued by some of the authors that organisms are complex systems and not just heaps of molecules, so that the analytical method does not suffice. Recent developments in systems theory offer the possibility to install a more comprehensive view of living systems what can be seen particularly in the field of evolutionary biology. It is true that any organismic activity is molecular, this is to say that it is based on molecular mechanisms. But it is also true that the whole organism displays certain patterns of behavior which are not just molecular. Any organism can be described as a system of different levels of organization – different levels of order and complexity – and it is important, therefore, to study all of the organizational levels and to see their peculiarities.

It should be obvious, however, that there is not one problem of reduction/reductionism, but that there are many problems linked together and that these problems appear at different levels of biological research and bio-philosophical reflections. The problems result from different philosophical interpretations of life and have a long and venerable tradition. Thus, in the first paper (Wuketits) a brief historical review will be found. This paper explains the main bio-philosophical traditions – vitalism and mechanism – and gives a brief survey of the meaning of the holistic, systems-theoretical perspective in today's biology. The paper by Hoyningen-Huene concentrates on epistemological reduction in the life sciences and discusses the arguments for and against (epistemological) reductionism. Ruse, having much sympathy for the sociobiological research program, offers a lengthy discussion of reductionistic strategies in sociobiology, whereas Löwenhard ponders on one of the central philosophical problems – the mind-body problem – from the point of view of contemporary neurobiology. In Mohr's paper you can find some clear statements concerning the problem of reductionism vs. holism in that field of biological research, which is frequently said to be the "most reductionistic": molecular. Wagner, last but not least, presents a systems-theoretical view of evolution which transgresses the orthodox (evolutionary) theory, i.e. the synthetic theory of evolution.

Thus, the main disciplines of modern biology are represented in this volume as far as the controversy reductionism vs. holism is concerned, so that we hope that the volume – as a whole – will clarify some of the central questions of bio-philosophy and that it will throw some light on the most intriguing bio-philosophical question: **What is life?**

Finally, we wish to express our thanks to the editors of the Theory and Decision Library for their continuing interest in this work, to the staff of the Kluwer Academic Publishers Group for the production of this volume, and last, but not least, to Manuela Delpos who installed the camera-ready copy of the typescript.

The editors

Franz M. Wuketits

ORGANISMS, VITAL FORCES, AND MACHINES: CLASSICAL CONTROVERSIES AND THE CONTEMPORARY DISCUSSION 'REDUCTIONISM VS. HOLISM'

> It seems that nature has taken pleasure in varying the same mechanism in an infinity of different ways.
>
> Denis Diderot

Paul A. Weiss was one of the founders of biological system theory. In the twenties, as a young student at the University of Vienna, he studied the behavior of butterflies and described animal behavior in general as a 'system reaction' (Weiss, 1925). This attempt has contrasted the reflex-chain theory of behaviorists. More than four decades later, in his splendid essay 'The Living System: Determinism Stratified', Weiss presented his attitude towards the living world by the following sentences:

> Biology must retain the courage of its own insights into living nature; for, after all, organisms are not just heaps of molecules. At least, I cannot bring myself to fell like one. Can You? If not, my essay may, at any rate, have given you some food for thought (Weiss, 1969, p. 400).

I have been very much attracted by this conclusion, and I admit that I actually cannot bring myself to feel like a heap of molecules, although I am aware that my body is indeed composed of molecules and that even my thinking and feeling might be traced back to molecular structures and functions.--

To be sure, the problem of reduction in the life sciences has often been treated in an emotional way; and the antithesis 'reductionism – holism' obviously has to do something with mere intuition. Gould (1980, p. 92) writes: "My intuition of wholeness probably reflects a biological truth." However, if organisms are reducible to inorganic compounds and if humans, too, are organisms – in fact, they are –,

3

P. Hoyningen-Huene and F. M. Wuketits (eds.), Reductionism and Systems Theory in the Life Sciences, 3–28.
© *1989 by Kluwer Academic Publishers.*

then any human being is reducible to molecules and atoms. Humans struggle against such conclusions – or, should I say that just their molecules are striving against these conclusions?

It is one thing to explain human life in terms of molecular biology, but it is another to feel like a heap of molecules. There is – or, at least, seems to be – a gap between scientific explanations of (human) life and human existential needs. The latter have sometimes influenced – and obscured – scientific discussions regarding human nature; and some scientists have tried to establish a picture of human life with resort to their personal beliefs and have thus confused science with religious or ideological claims. (A recent example is Sir John Eccles` The Human Mystery, 1979.) But many biologists beyond any religious traditions have no less complained about the reductionist attitude towards life and have called for additional modes of explanation in biology. Simpson (1963) for one has demanded a `second form` of explanation, namely a `compositionist explanation` which should help us to explain the purpose of structures and functions in relation to the whole organism.

At any rate, the problem of reduction/reductionism has given rise to many controversies. The discussion `reductionism vs. holism` is an integral part of philosophy of biology (see Sattler, 1986); it is an important discussion within and outside the biological sciences, for, not least, it reflects the old and venerable question what is life?

In this paper I shall give, first of all, a brief survey of different interpretations of living phenomena (vitalism, mechanism) and, then, a review of the present situation. Whereas some of the other papers in the present volume deal with special aspects of reduction (reductionism) and/or holism, i.e. the systems view, in particular areas of biological research (Löwenhard, Mohr, Ruse, Wagner), I shall give a more general historical account. History of science often serves as an `eye-opener` and helps us to understand the current controversies.

WHAT IS LIFE?

For some scholars the phenomena of life are at variance with the laws of physics and chemistry, others try to substantiate the thesis that life is nothing else but physics and chemistry. At least, this has been the situation in many discussions in the field of controversy between the advocates of the animist–vitalist tradition and the defenders of mechanistic theories. (I shall characterize the vitalistic and the mechanistic interpretations of life in the next two paragraphs.)

Certainly, there are only a few biologists today advocating a strict vitalism. However, there are many standing in contrast to a strict mechanism. Perhaps it is true that "gradually the discussion between biologists changed and the opposition of mechanism vs. vitalism gave way to that of reductionism vs. integrationism" (de Klerk, 1979, p. 9). But I am not sure whether or not all biologists have abandoned their belief in the existence of specific vital forces, because, as I mentioned above, explanations of life have frequently been influenced by religious claims. Nevertheless, it is possible for a biologist today to discuss the crucial question 'what is life?' beyond the old mechanism-vitalism controversy. Looking at living systems, a biologist can notice some essential features common to all organisms – protists, plants, animals (humans included); he or she is not necessarily affected by ontological questions and premises. For example, Mohr (1965) has described six characteristics of living systems which apply to all organisms and thus have the quality of 'universal biological laws', i.e. biological principles (see table I; see also Mohr, 1977; Wuketits, 1978, 1983, 1985). Mohr (1977, p. 62) writes:

> Universal laws in biology are principles that are valid for all living systems (but not necessarily for all physical systems). A biological principle that is actually a consequence of the second law of thermodynamics as applied to open systems can be expressed in symbolic language aG $\neq$ O whereby G is the symbol for ... free energy. This principle is valid for every open system, living or nonliving. Since all living systems are open systems, it is valid for all living systems.

Such statements amount to the assertion that physical (or chemical) principles/laws are indeed valid for living systems, but that these systems, at the same time, transgress the principles/laws of physics (and chemistry).

TABLE I

Principle of development	Every living system is a mutable system, undergoing processes of structural and functional change
PrinciplG ≠ 0	Every organism continuously requires the supply of free energy to compensate for the continuous production of entropy (i.e. living systems are not in thermodynamic equilibrium)*
Principle of compartmentation	Every organism is compartmented, showing different structures (the structures of an organism themselves are hierarchically organized)
Principle of regulation	All living systems are able to maintain their structures and functions in spite of environmental fluctuations ('self-restoring')
Principle of heredity	The transmission of genetic information is a feature of all organisms; the mechanism by which a sequence of nucleotids is translated into a sequence of amino acids, is the same in every organism
Principle of catalysis	On molecular level the activities of all organisms are regulated by enzymes (biochemical catalysts)

* As will be remembered, the insight that organisms are negentropic systems, so to speak, i.e. systems producing negentropy ('order'), was formulated by the physicist Erwin Schrödinger (see Schrödinger, 1967).

Life is a particular phenomenon in the universe; therefore, biological principles are special principles and living systems rather sophisticated material entities. Consequently, if you look at biological textbooks you will find words like the following:

> Life is ... a very special phenomenon. We believe that it is the behavior pattern which matter exhibits when it reaches a certain level of organization and complexity ... We can observe from a distance, and disassemble one part of the whole for study, whether the part is a population, an individual, an organ or a molecule. We can propose ideas about how living systems work and why they exist as they do, and check the simpler of these ideas experimentally (Wilson et al., 1973, p. 10).

All living systems include at least two characteristics that cannot be found in physical or chemical systems:

(i) There is a certain **individuality** proper to organisms.

(ii) Because of their individuality living phenomena are not reproducible artificially.

Mathematical symbolism, therefore, cannot be generalized to all processes of life; and it is naive to believe that biological problems are to be solved by applying mathematical formulas. To be sure, formal tools can help us to specify the one or the other problem in biology, but they do not help us to understand the essential features of organic evolution and development. I have objections to Bunge's claim for using formal tools in the life sciences and in the philosophy of biology (see Bunge, 1979), not because I am not prepared to appreciate the meaning of mathematics, but because it is simply not true that the philosophy of biology, as Bunge (1979) believes, will reach its maturity only by our using of formal tools.

Things might get clearer if you remember what has been known under the concept of **Gestalt**. Humans perceive any living being as a certain **Gestalt**, and the same is true to the perception of other animals. The concept **Gestalt** has a venerable tradition, sometimes, however, it is linked to idealistic philosophy. To be sure, this concept is not necessarily an idealistic one. Our perceiving apparatus calculates, so to speak, the `figure` of a living being preconsciously, that is to say that the process of perceiving the world around us immediately gives us a true notion of whether an object is a living or a non-living one. This is the great performance of Gestalt-perception which operates without being grasped at the conscious level (see Lorenz, 1977) and which, then, helps us to distinguish between living beings and other objects at a prerational level (see Wuketits, 1985). Thus, in a nutshell, when we are looking at the world around us we immediately recognize a system

either as living or as non-living, and we notice that "most systems are either
obviously living or **obviously** dead" (Marquand, 1968, p.12, my italics).

Defining life, therefore, is indeed – at least to a certain extent – a matter of
intuition. But it is true that by mere intuition we cannot get sufficient scientific
explanations and definitions. Our unaided vision does not suffice if we want to
know more about the hereditary principles of reproduction, about biochemical
catalysis, and so on. Hence, the six biological principles (table I) are not the
result of intuition, but rather the result of rational scientific research. Within this
research and within our rational reflections upon **life** some crucial questions
emerged: `What is the essence of life?` `What makes living systems **living**?` `Do we
need special vital forces to explain the phenomena of life?` A biologist today may
be convinced that such questions are nonsense and that it is better just to
characterize living beings instead of reflecting upon `life` in an abstract sense.
But since life is also a philosophical problem and since the problem of life is one
of the most intriguing questions of human thinking, this, I feel, would be an
incorrect stance.

Let me give a brief outline of some doctrines which up to now have influenced the
discussions of life at the interface between biology and philosophy.

THE VITALISTS` CREDO

According to vitalism life and all its particular expressions depend on specific
agents which are not to be explained in terms of physics and chemistry. More than
this, some advocates of this doctrine (e.g. Driesch, 1928) made the claim that a
spiritual principle is acting upon any living being. In other words:

> Vitalistic positions assume in some form or other the
> existence of an agent which actively selects and arranges
> matter in the organism. Some vitalist approaches assume
> furthermore that this agent, which may be a rational soul,
> can exist separately from matter and that the organism
> is in a healthy functional state so long as the vital agent
> remains in control (Lenoir, 1982, p.9).

Thus we have to distinguish between, at least, two kinds of vitalism (full details
in Wuketits, 1985): **animism** and a type of **naturalistic vitalism**. Let me explain.

Animism or, as Kochanski (1979) puts it, `psychovitalism` is a metaphysical
approach to an understanding of life. The animists claim that living phenomena
depend upon spiritual principles, immaterial forces which cannot be defined in
scientific terms. The central tenet of animists has been that such forces act

teleologically. According to this view all living phenomena would be directed towards a certain goal. This idea was expressed, to name but a few, by Galenos (129–199), Paracelsus (1493–1541), and Swedenborg (1688–1772). In the twentieth century particularly Driesch (1867–1941) advocated this idea.

The other type of vitalism has been less speculative than animism, for its advocates have not regarded the `life force` as a mystic and unintelligible factor, not as a soul or something like that. The main concern of some naturalists in the eighteenth and in the nineteenth century was rather to conceive vital forces as agents acting in – and not coming from outside – the organism. The meaning of the **Lebenskraft** as conceived, for example, by the German physiologist Müller (1801–1858), is not that of an immaterial principle. However, it was coined to characterize organic laws transgressing the range of physical explanations. The same is true to the vitalistic conceptions of scholars like Haller (1708–1777) and Kielmeyer (1765–1845). And Blumenbach (1752–1840) worked out, with some insight, emergent properties of life based upon a specific order of elements in the living organism. This naturalistic type of vitalism has been called `vital materialism` (Lenoir, 1982), for its proponents did not believe in supranatural factors but only wanted to underline the specifity of living beings among all objects in the universe. Blumenbach`s notion of the **Bildungstrieb,** for instance, reads as follows:

> The term **Bildungstrieb** just like all other **Lebenskräfte** such as sensibility and irritability explains nothing itself, rather it is intended to designate a particular force whose constant effect is to be recognized from the phenomena of experience, but whose cause, just like the causes of all other universally recognized natural forces remains for us an occult quality. That does not hinder us in any way whatsoever … from attempting to investigate the effects of this force through empirical observations and bring them under general laws (Blumenbach, 1797, quoted by Lenoir, 1982, p.21).

(The concept **Bildungstrieb** means something like a developmental force`; however, it is difficult to give an adequate English translation.) Likewise Kant, with regard to purpose in living nature, wrote in his **Critique of Judgement** (1790, see 1951, p. 219) that

> the first principle required for the notion of an object conceived as a natural purpose is that the parts, with respect to both form and being, are only possible through their relationship to the whole .. Secondly, it is required that the parts bind themselves mutually into the unity of a whole in such a way that they are mutually cause and effect of another.

Hence, we can see that purpose in living nature does not necessarily means the existence of a spiritual principle.

Yet the proponents of vitalism - be it in the strict sense of the animist tradition or in the sense of a vital materialism - have been convinced that there is a special vital principle common to all living beings and that this principle cannot be explained in terms of physics or chemistry. The vital principle has been expressed in many different concepts, e.g. entelechy, **spiritus**, **anima**, **Lebenskraft**, vital power, **élan vital**, and others (full details in Wuketits, 1985). None of these concepts, however, really explains the specifity of living systems. To say, for example, that there is an **élan vital** managing the activities of organisms (Bergson, 1907) means just to confess that there is a peculiar (non-mechanical) drive; but nobody actually knows what this drive means. No wonder that biologists standing in the evolutionist tradition have dismissed any kind of vitalism "Vitalism has been excluded from science because it does not meet the requirements of a scientific hypothesis" (Dobzhansky et al., 1977, p. 488); and it has been refused because now there are evolutionary explanations available to cope with phenomena like purpose (see below). However, a few authors even most recently have advocated a vitalist position. An example is Portmann (1965, 1973) who has claimed that organisms are to be characterized by their **Selbstdarstellung** (`display` or, in the literal translation, `self-portrait`) and by their **Innerlichkeit** (`centricity`, `inwardness`). (See on Portmann's biophilosophy Grene, 1974.)

This brief comment on vitalism may suffice in the present context. It should be clear that the vitalists` propositions are untenable in the light of modern biological research. But one should be aware that, despite the shortcomings of vitalism, there is a grain of truth in this doctrine. This is the idea that organisms are not just machine-like systems to be explained in physical terms. This claim, however, has been made by the advocates of the mechanistic tradition.

THE MECHANISTS` CREDO

In contrast to vitalism the mechanists have based their doctrine on the assertion that organisms are in fact something like machines, functioning mechanically, performing their activities like mechanical systems. In its most radical version mechanism even amounts to the claim that living systems are nothing else but machines. Such a `machine theory of life` was defended by the French physician and philosopher La Mettrie (1709-1751).

More than two thousand years ago a school grew up which can be characterized by the assertion that living systems can be explained mechanically, i.e. in terms of mechanics. This school was established as a reaction to Aristotle's biology. Aristotle himself was not an animist, but what can be seen in his concepts of formal and final cause (**causa formalis, causa finalis**) is a kind of naturalistic vitalism. The first mechanistic conceptions of nature were founded, however, before Aristotle, namely by the **atomists**. Aristotle's philosophy of nature was a non-mechanistic interval, so to speak, because his pupil Theophrast soon established a mechanistic interpretation of life.

Anyway, one should distinguish between ancient ideas of mechanism and the machine theory of life which resulted from some philosophers' approaches to nature in the seventeenth and in the eighteenth century and which has been a strict mechanical world view. The materialists and mechanists in ancient Greek philosophy had not drawn analogies between organisms and machines (see, e.g., Grmek, 1972). The scene changed when humans were going to develop more complex technologies and automata. However, the 'mechanization' of living nature in modern times has been the result of two research programs and strategies (see, e.g., Wuketits, 1985; see also Taylor's highly readable and well illustrated **The Science of Life**):

(i) In the seventeenth century **iatromechanics** was established. The founders of this school, above all Borelli (1608–1679), studied the principles of locomotion in animals and humans and described locomotion in mechanical terms using Galileo's principles. Thus, physiological problems were studied as problems of physics (mechanics) and some biologists argued that organisms are to be regarded as machine-like systems.

(ii) The next step was the conclusion that organisms are **nothing else but** machines. This conclusion was most dramatically formulated by La Mettrie in his controversial **L'Homme machine** (1848). There seemed to be some substance to this conclusion because of the possibility to build complex machines somewhat similar to organisms.

With respect to the second point it might be worthwhile to give some examples.

In the seventeenth century the French engineer Vaucanson (1709–1782) built some automata which clearly demonstrated man's technical ability to control physical laws and to translate, as it were, the knowledge of these laws into machines. Among Vaucanson's marvellous machines was the 'mechanical duck', an automatic 'duck' able to perform different activities like eating, moving its wings, and so on. This 'duck' and, generally, any of Vaucanson's machines "demonstrated, or better, gave an illusion of demonstration, that man could mechanically depict living processes" (Grmek, 1972, p. 194; see on this also Cohen, 1966). Understandably, the existence of such machines influenced the interpretation of life. Grmek (1972, p. 194) has made the following point:

> Vaucanson called his automata 'moving anatomies' and actually he preceded his technical achievements by a detailed study of anatomy and physiology. His duck (constructed about 1733) was able to digest, that is to swallow grains and, a little later, to eliminate some excrement-like matter. It was, in fact, a dishonest technical trick and not an imitation of digestion, but for us it is significant that so many people accepted Vaucanson's affirmations as a real solution of a complex physiological problem.

More than this: For many people Vaucanson's automata did not only solve physiological problems, but through the construction of such machines like the mechanical duck the differences between living and technical systems seemed to be eliminated. And the theoretical, philosophical conclusion simply was that organisms are just machines and nothing else.

Another kind of mechanisms, however, emerged in the nineteenth century and is due to Darwin's theory of **natural selection**. From the point of view of this theory organisms are not machines but systems which depend upon – and are changed by – a mechanical principle, the principle of natural selection operating by the physical environment of any living system: The organism is dominated by external selective agencies. Darwin's theory stands in contrast to any vitalist claim and has ruined the notion of a universal teleology (see next paragraph).

Finally, in the twentieth century a new type of mechanistic interpretations of life has been developed. I call this 'molecular mechanism' bearing in mind that it was strongly influenced by models in **molecular biology**. The lesson which molecular biology has indeed taught us is that all living systems are composed of molecules and that molecules are responsible for an organism's structures and functions, and for the processing of genetic information. Thus, there grew up – and has endured – **molecular biophysics** as a discipline "centred mostly on those problems related to the structure and physical properties of proteins and nucleic acids, which determine their biological function" (Vol'kenshtein, 1974, p. 9). I shall refrain from going into details here, because Mohr's paper (in this volume) is devoted to molecular biology and the problems of reductionism. However, I am probably right that the enormous success of molecular biology has tempted many contemporary biologists to look at living systems only by studying their molecular compounds and processes at the molecular level.

TABLE II

(i) Greek atomism, 5th to 3rd century B.C. There is no fundamental difference between living and non-living objects (Democritus); the 'qualities of life' are reducible to universal laws of matter

(ii) Machine theory of life, 17th and 18th century. (a) Iatromechanics: Activities of living systems are explainable in terms of mechanics (e.g. Borelli) (b) Machine theory s. str.: Organisms are nothing else but machines (La Mettrie)

(iii) Naturalisms.str., 19th and 20th century. The Evolution of life is based on mechanical agents which are defined by the environment (external selection): Theory of natural selection (Darwin) and other theories in the late 19th and in the 20th century, as far as they are darwinistic

(iv) 'Molecular mechanism' (mechanistic conception of life based on molecular biology) Organisms are reducible to their molecular compounds: organisms are nothing else but heaps of molecules

---

To summarize briefly to this point, I have argued that there have been different types of mechanism. Table II shows their basic claims. However, every advocate of the one or the other type of mechanism has been convinced that living systems are physical/chemical phenomena or that they are, at least, influenced by such phenomena and that, finally, there is no reason to believe in particular vital forces or even spiritual principles.

DARWIN'S THEORY – THE END OF TELEOLOGY?

In his celebrated On the Origin of Species Darwin wrote: "From the war of nature, from famine and death ... the production of the higher animals directly follows" (Darwin, 1859, see 1958, p. 450). Moreover, he remarked "that, whilst this planet has gone cycling on according to the fixed law of gravity, from so simple a beginning endless forms most beautiful and most wonderful have been, are being evolved" (Darwin, 1859, see 1958, p. 450). As I have stated previously, Darwin's theory was a mechanistic theory of life. The `war of nature` has been conceived by Darwin – although he used such phrases as mere metaphors – as a natural principle, a mechanistic principle that would favour the well-adapted, the fittest, and eliminate all the other organisms.

Darwin dismissed the idea of a `creative factor` in nature, and he refused the notion of teleology in the living world (see, e.g., Himmelfarb, 1968; Hull, 1974; Lenoir, 1982). Was this, then, the end of an old controversy? To be sure, it was not. Remember that vitalistic and teleological thinking has endured up to the present and that in the twentieth century there have been some biologists advocating the vitalist tradition.But it is true that a great majority of today's biologists have not much sympathy for vitalism. Now, what's the matter with teleology, the idea of plan and purpose in (living) nature? Let me state at once that mechanism is not necessarily at variance with teleological thinking. Particularly, the machine theory of life may include the notion of teleology: An organism may be a machine, but this machine might have been created by God and built according to a divine plan; hence, the machine could serve a `higher purpose`. And, indeed, the advocates of such a machine theory or, generally, mechanism have not necessarily excluded the notion of a universal teleology (an exception, of course, was La Mettrie). However, the scene changed in the after math of Darwin's theory. Those who have taken this theory seriously – and this has been a majority of biologists – have discarded the concept of teleology and, consequently, dismissed the idea that man is the apex of creation. Simpson(1963, p. 12–3), for one, bluntly stated that

> in the world of Darwin man has no special status other
> than his definition as a distinct species of animals: He is
> in the fullest sense a part of nature and not apart from
> it. He is akin, not figuratively but literally, to every
> living thing, be it an amoeba, a tapeworm, a flea, a
> seaweed, an oak tree, or a monkey – even though degrees
> of relationship are different and we may feel less empathy
> for forty-second cousins like a tapeworms, than for,
> comparatively speaking, brothers like the monkeys. This is
> togetherness and brotherhood with a vengeance, beyond
> the wildest dreams of copy writers of theologians.

I think that most biologists today would agree to such conclusions. But although a biologist today would hardly believe in a cosmic teleology, he or she does not refrain from using teleological concepts in a particular sense. It is true that "wherever there is reference to part–whole relations (in morphology), to means–end relations (in physiology), or to goal–governed processes (in embryology), there is teleological discourse" (Grene, 1974, p. 174). In fact, living systems are organized teleologically in a special sense, they perform activities for the sake of survival. But these activities do not mean that there is a vital force or something like that. According to Darwin's theory of evolution in general, all purposeful structures/functions in living nature are to be explained with resort to an optimization of the genetic program which itself is a result of evolutionary processes. Therefore, the teleology of modern biology is not to be confused with the classical notion of teleology expressing the view that organisms serve a cosmic goal. In order to avoid misunderstandings, many biologists and philosophers of biology today use the term **teleonomy** instead of teleology (to name but a few, Dobzhansky et al., 1977; Hull, 1974; Lorenz, 1977; Mayr, 1974; Mohr, 1977; Wuketits, 1980,1983,1985). Mayr (1974), for instance, says that teleonomy or a teleonomic explanation does not give comfort to advocates of classical teleology or vitalism. But this is clear, because any teleonomic explanation refers to the fact that all organisms are the outcome of evolution and that any living system, with regard to its structure and behavior, is developing according to a genetically stabilized program. Thus, in a previous paper I wrote:

> A process is said to be teleonomic, if it relies upon the efficiency of a program and therefore implies the property of goal-directedness. As a matter of fact there are two characteristical components proper to teleonomic events: On the one hand their regulation based on the program, on the other hand their dependence upon the existence of a certain goal (e.g. a physiological function, the achievement of a new geographic position etc.) (Wuketits, 1980, p. 285).

It is indeed true that, if you look at the development of an organism (embryology), there is goal-directedness. "Development", writes Jacobs (1986, p. 392), "is not mere alteration or change. It is an end-oriented process. But biological development and ends can be empirically specified. They are not immaterial processes." Of course, they are not – they are the result of encoding a genetical program which has been develops in the course of evolution. Hence, if a biologist is using teleological terms, he or she does not necessarily believe in vital forces.

The question **what for?** is one of the standard questions in the life sciences. But a biologist who takes evolution seriously will not answer this question with resort to obscure (vital, spiritual, divine) principles. He or she will agree to Lorenz (1965, p. 24) that

> if a biologist says that the cat has crooked, pointed claws 'with which to catch mice', he is not professing a belief in a mystical teleology, but succinctly stating that catching mice is the function whose selection pressure caused the evolution of that particular form of claws.

In short, biologists cannot refrain from using teleological phrases, but these phrases ('what for?', 'for the sake of') only mean that a living being is able to perform certain activities on the basis of its genetical program and 'for the sake of'its survival. 'Survival', however, is nothing supranatural, it is intrinsic to living systems. Thus, it is true that Darwin's theory has been the end of teleological speculations if teleology is meant to be a supranatural principle or to depend upon such a principle.

But can mechanism give us a proper understanding of living systems and their teleonomic organization? Many biologists believe that it cannot and that a view of organic beings is needed which is going beyond the mechanists' credo but which, at the same time, is not to be confused with the vitalists' doctrine.

THE HOLISTIC VIEW OF LIFE – BEYOND VITALISM AND MECHANISM

I has become apparent that vitalism does not meet the requirements of scientific explanation and that mechanism does not offer a true insight into the specifity of living beings. Perhaps "the struggle between mechanism and vitalism has brought science no nearer to an explanation of the 'essence of life'" (Cassirer, 1950, p. 216). But this struggle prompted biologists and philosophers of biology to reflect the status of the life sciences, and "it has compelled biology again and again to examine the question concerning **its own nature as a science**" (Cassirer, 1950, p. 216). Some biologists – and philosophers of biology – who have examined this crucial question, attained to the conclusion that an 'organism-centred' view of life is necessary and that neither mechanism nor vitalism are apt to account for the 'true nature'of living beings. Thus **holism** emerged as a new paradigm in biology and biophilosophy.

The situation in the life sciences in the nineteenth and at the beginning of the twentieth century was this: On the one side there was 'themechanists' claim that organisms can be sufficiently explained as mechanical systems; on the other side the vitalists made use of obscure principles like entelechy, **élan vital**, and so on; it was a rather unpleasant situation. Biologists were in the quandary either to refer to vital forces at the cost of acceptable scientific explanations, or to ignore phenomena that characterize the very nature of living systems (self-regulation, self-maintenance) at the cost of the insight into essential features in the organization of living beings. What was needed, then, was a theory of living matter going beyond vitalism and mechanism and beyond a metaphysical conception of wholeness (**Ganzheitsschau**) and its antithesis, namely atomism. The zoologist Ritter made a decisive step by proposing the thesis that "the organism in its totality is as essential to an explanation of its elements as its elements are to an explanation of the organism" (quoted by Beckner, 1967, p. 549). Ritter called his view 'organismal'('organismalism'); the term was then replaced by the concept **organismic (organicism)**.

It was through the work of Paul A. Weiss and – even more – the writings of Ludwig von Bertalanffy that the organismic view of life became a fully fledged natural philosophy. However, a distinction should be made between this kind of holism, organicism, and the metaphysical holistic views. The latter was advocated by scholars like Meyer-Abich (1934) who felt bound to idealistic philosophy and claimed that the wholeness is more important than its parts and that it could be studied without resort to the parts. This has been a speculative metaphysical view. Its advocates have indeed been aware that any organism consists of parts (organs, cells) but they have insisted that the parts are determined by the whole. Needless to say that this kind of holism is an obsolete style of thinking and looking at living systems. In contrast to this a systems-theoretically founded version of holism is of great importance in biology and philosophy of biology. The central claim of this holism is that neither the whole determines its parts nor the parts determine the whole but that a complex interaction between parts and whole is to be supposed, so that there is determination in two directions. Such a view has been developed by Bertalanffy in his **General System Theory** (1968) which strongly has influenced biology as well as other sciences (see Laszlo, 1972). Bertalanffy's system theory can be summarized as follows:

(i) The whole (of an organism) is more than the sum of its parts. This might be a truism (Beckner, 1967), but I feel that biologists often forget that studying molecular biology does not mean to adopt the full wisdom of biology.

(ii) Living beings are open systems. Physics traditionally deals with closed systems, so that there has been a gulf between physics and biology. Modern physics, however, includes open systems too; this is particularly true to **nonequilibrium thermodynamics** (Prigogine, 1972) which has shed some light on open systems in biology (see also Leinfellner, 1988).

(iii) Living systems are not static systems; they are to be regarded as continuous processes. In Bertalanffy's words: "This continuous decay and synthesis is so regulated that the cell and organism are maintained approximately constant in a so-called steady state (**Fließgleichgewicht**)" (Bertalanffy, 1968, p. 165).

(iv) Organisms are homeostatic systems. This insight was influenced by Norbert Wiener's cybernetical approach and by information theory; it amounts to the assertion that any living system represents a dynamical interplay at all levels of its organization.

(v) Organisms 'arehierarchically organized systems. Any organism is structured in a way so that its individual members (organs, cells) are 'super-systems' of other elements or levels of organization. "Such superposition of systems is called **hierarchical order**. For its individual levels, again the aspects of wholeness and summativity ... apply" (Bertalanffy, 1968, p. 74).

With respect to hierarchical organization the following points are pertinent.

The concept of hierarchy, when put into a systems-theoretical framework, is not to be confused with the classical notion of the scala naturae which was a statical conception. What we actually find in living systems is a dynamical hierarchy based upon constant interactions between the levels of organization. "Hierarchic patterns are advanced **patterns of interdependence** in the sense that the dependent structures ... are mutually subordinated to form ranks, grades, or classes" (Riedl, 1977, p. 354, my italics). Generally, we may suppose that the interactive relations among hierarchically organized systems are the dynamic components of nature, the intrinsic characteristics of the emerging web of the world (Wuketits, 1982). It might be that some people feel that this is mere phraseology. But perhaps those people will get the point when they consider the following illustration given by Paul A. Weiss:

> To dramatize the need for viewing living organisms as hierarchically ordered systems, I shall give you the following facts to ponder. The average cell in your body consists to 80 % of water and for the rest contains about 10^5 macromolecules. Your brain alone contains about 10^{10} cells, hence about 10^{15} ... macromolecules (these figures may be off by one order of magnitude in either direction). Could you actually believe that such an astronomic number of elements, shuffled around as we have demonstrated in our cell studies ..., could ever guarantee to you your sense of identity and constancy in life without this constancy being insured by a superordinated principle of integration? (Weiss, 1969, p. 370).

The concept of dynamical hierarchy in organic systems has thrown new light also to another important problem, namely biological **causality**. It has become apparent that causality in living systems is more complex than the classical notion of linear cause-effect relations.

The classical notion of causality was, some way or other, **deterministic**; it suggested that events at lower levels of organismic organization cause events at higher levels. However, it seems to be clear that this upward causation, as it were, tells us only the half of the story and that we have to suppose what Campbell (1974) called **downward causation**. This is to say that, if you see organismic systems or other complex systems as hierarchically organized systems, you have to take into account a flux of cause and effect in two directions, up and down the scale. Thus, organisms appear as complex networks of causal relations and the old one-way-causal paradigm is to be replaced by the notion of **network causality** (Riedl 1977, 1979; Wuketits, 1981).

Furthermore, you will easily recognize that the holistic, systems-theoretical view is at variance with **ontological reductionism**, according to which all natural phenomena, particularly organisms, are nothing else but heaps of smaller, less complex parts. In contrast to this view we have to assume that, as mentioned above, the whole is more than its parts and that it results from complex **interactions** between elements at different levels of organization. Organisms are indeed to be characterized by increasing complexity, by **fundamental complexity**, "brought about by increase of metabolism", as Cramer (1979, p. 138) has emphasized, and consisting of "indescribable decomposition and formation of structures".

To be sure, any biologist working in a special branch of research will be well advised to separate smaller parts from the whole system. This will be – and actual is – a research strategy, a methodological rule. An anatomist, for instance, has to decompose the organism if he or she wants to know something about its composition, i.e. its different organs. But the function of the different organs can only be understood within the whole organism. Hence, what is needed is not only the **analytical** method, but also the **synthesis** (Wuketits, 1987), the view regarding the whole organism or, as Simpson(1963) puts it, **compositionism**. Both analysis and synthesis are integral parts of biological thinking.

This brief comment on reductions may suffice in the present context, for Ruse (in the volume) gives further details concerning reduction/reductionism in the life sciences.

In short, organismic biology (organicism, holism) has given us a broader philosophical framework to the understanding of living systems; it is going beyond mechanism and vitalism which have been one-sided views of organic nature and which have either simplified (mechanism) or obscured (vitalism) the very nature of life. In the twentieth century many biologists have taken a holistic view, and they have been accompanied by some physicists (e.g. Polanyi, 1968). (For more details see Kochanski, 1979.) Since holism – as a systems–theoretical perspective – is not to be confused with a mere idealistic claim (**Ganzheitsschau**), it represents a scientific paradigm which can be applied to the study of **interrelations** between different elements that constitute the emerging properties of complex systems. (See also Bunge, 1977.)

BEYOND DARWINISM? – SOME REMARKS ON HOLISM AND EVOLUTION

The holistic (systems–theoretical) approach to living systems has given us a deeper insight into the processes of organic evolution. Darwin's theory of natural selection has indeed proved to be true, but, as a matter of historical fact, his ideas about genetic mechanisms of evolution were rather vague. Classical genetics had to be rediscovered at the beginning of the twentieth century. It was through the work of many naturalists that we could get some insight into the genetical composition of organisms and the basic principles and mechanisms of the hereditary processes. However, **genes**, the basic units of heredity, were regarded as entities existing independent of each other; moreover, it was suggested that any gene determines one particular character of an organism. Thus, the one-gene-one-character conception was developed, a conception that clearly relies upon the one-way-causal paradigm. Therefore, it was an important insight that a "genotype consists of a number of co-adapted gene complexes and that even the gene pool of a population as a whole is well integrated and co-adapted" (Mayr, 1975, p. 381). To be honest, one should admit that we have not yet really understood the genetical systems in higher organisms and that in this area of research much work is still to be done. But the notion of the **epigenetic system**, i.e. the system of gene-interactions (Riedl, 1979) probably reflects the truth. This notion gives us to understand genes as complex interacting elements, linked together and linked with the phenotype of an organism like a network (remember, once more, the concept of network causality).

As I have stated previously, Darwin's concept of natural selection has been a mechanistic concept. Selection has been regarded as a mechanism working from outside the living system, so that, in a way, it would be moulded by its environment. Thus, any living system has been said to be the result of external (selective) agencies. But is this view apt to explain and to understand the immense order and complexity of organisms at all levels of their organization? **Feed-back causality** helps us to come nearer to the truth. Selection has indeed been the major factor creating organisms and directing the evolutionary change, but we have not only to consider external (environmental) selection but also **internal** selective constraints. Riedl (1977, p. 363) writes.

> While mutation and environmental selection are "blind" and "short-sighted", this proposed third factor of evolution is far-sighted, although it can only look backwards, into the responses within its own organization. It operates ... by selection, but the requirements of this selection differ from those of environmental selection, in the same manner as the patterns of organization within an industrial plant differ from the patterns of opinions and desires in the market.

This means that organic evolution is not only determined by external forces, but also by **organismic constraints** (see also Gould and Lewontin, 1979). Besides, the concept of **feedback-selection** (Wagner, 1981, and in the present volume) is of some importance. It is to be supposed that there is a backflow of information from the phenotype to the genotype so that the epigenetic system, in the long run, copies the functional order of its own product. That means, to put it metaphorically, that the epigenetic system "must ... not wait for the market in order to be told about every possible mistake. Most of the possible mistakes are taken care of in advance by the interior management of the system itself" (Riedl, 1982, p. 47). Lamarck was perhaps not completely wrong, although many of his assertions have been refuted. From the point of view of a systems conception of evolution both Lamarckism and Darwinism contain some truth.

Lamarckists would suppose that there is a flow of biological information from the phenotype (P) directly to the genotype (G) and that, therefore, adaptations acquired during the lifetime of any single organism are transmitted to the next generation:

flow of information

G ⇐------------------------------- P

The advocates of the Darwinian view, on the contrary, assume that there is only a flow of information from the genotype to the phenotype:

flow of information

G -----------------------------⇒ P

From the point of view of a systems theory of evolution, however, we maintain that there is flow of information in both directions:

flow of information

G ⇐---------------------------⇒ P

The basic idea underlying this model is that living systems represent a complex network of causal relations.

I am aware that these propositions would require further elaboration. But I feel that the reader, even if he or she is not familiar with modern biological theorizing, will be able to see that a systems-theoretical, a holistic view of evolution challenges many classical and established concepts.

LIFE AND THE AUTONOMY OF BIOLOGY AS A SCIENCE

Ayala (1972, p. 6) clearly stated:

> A majority of biological problems cannot be as yet approached at the molecular level. Biological research must then continue at the different levels of integration of the living world, according to the laws and theories developed for each order of complexity. The study of the molecular structures of organisms must be accompanied by research at the levels of the cell, the organ, the individual, the population, the species, the community, the eco-system. These levels of integration are not independent of each other.

Such sentences have still their importance. Many phenomena of life cannot be sufficiently explained in terms of molecular biology, and if we really want to understand the complexity of the biosphere we have to refer to different levels of organization at and above the level of an individual organism.

If, now, one accepts that living beings are multi-level-systems of increasing complexity displaying specific characters, one might also accept that the scientific study of these systems, i.e. biology is not reducible to other sciences (physics, chemistry). To be sure, living systems are not at variance with fundamental physical and chemical principles, for they are, in fact, **material** systems and not systems governed by some spiritual principles. But the physical and chemical laws represent only the **necessary**, and not the **sufficient** conditions and constraints of living beings. The development of life transgresses the domain of physical (and chemical) sciences and entails principles which are not to be discovered at the physical (and chemical) level.

The specific arrangement of material elements in living systems and the patterns of evolution and development of life requires specific questions and answers, that is to say particular **types** of questions and answers. Thus, the question **what for?** is a typical biological question, it is unknown to the physicist. consequently, teleological explanations, i.e. explanations of the teleonomic organization of organisms (see above), are not necessary in the physical sciences, but they are indispensable to biology. "Teleological explanations, then, are distinctive of biology among all the natural sciences" (Ayala, 1972, p. 15) and it is for this reason that biology is to be regarded as an autonomous science (see also Wuketits, 1978, 1983). Elsasser might be on the right track in trying to establish a biological theory that would be compatible with quantum mechanics (see Elsasser, 1981). I mean that of course one can look for general explanations in science. But if the result, then, is that all natural systems (including living systems) are to be explained in terms of quantum mechanics and to be reduced to quantum mechanics, then something is wrong. Any approach to reduce living beings to micromechanical processes and mechanisms is comparable with the old machine theory of life. "The flaw lies", as Weiss (1969, p. 385) pointed out, "in equating science with the doctrine of micro-precise causality, or ... `micro-determinism`."

In short, physical (and microphysical) laws are indeed necessary conditions to life, but they are not sufficient to explain and to understand the peculiarity of living systems. Likewise biological laws and principles are necessary but not sufficient conditions to explain the particular pathways of human cultural systems. Thus, in order to understand the emerging complexity of nature, an approach is necessary which takes into account different levels of organization. It is true that

> the laws which govern (biological) combinations certainly cannot contravene the laws of physics and chemistry. But over and above the laws of physics and chemistry the biologist can look for further principles which might help us to understand why, from the potential universe of what is permitted by physics and chemistry, certain combinations do occur and others do not (Barker, 1976, p. 378).

Moreover:

> When units are arranged together in particular ways it sometimes happens that the whole which they form has certain properties which are not present in the units themselves or even in different combinations of those units. Such properties are said to 'emerge' at this level of organization (Barker, 1976, p. 381).

CONCLUDING REMARKS

At the beginning of this paper I have remarked that the controversy **reductionism** vs. **holism** in biology is an emotionally obscured topic. However, I am convinced that it would be better to deal with this problem beyond any emotional stance. Certainly, humans themselves cannot be excluded from the debate; and nobody, I suppose, actually feels like a heap of molecules – and nothing else. But at the same time we should be aware that we all are natural systems and, not withstanding the social and cultural constraints to our behavior, systems that have emerged in the course of organic evolution; and that even our cognition and knowledge is a product of (organic) evolution (see on this, e.g., the contributions in Wuketits, 1984). However, appreciating the organismic, holistic (systems–theoretical) view, we can conclude that any ontological reductionism is desperately defective.

In the present paper my intention was just to focus the reader's attention on some old problems in biology and the biology of philosophy, problems which are still alive. Both mechanism and vitalism have been defective approaches to an understanding of life. Only the holistic, systems-theoretical view seems to offer a promising perspective. Studying the lower levels in the organization of living systems, i.e. studying molecules, has become indispensable for biology; but this cannot mean that biology, now, is (or should be) nothing else but molecular biology. Biology as the study of life contains both molecules and organisms. Organisms are indeed composed by molecules – but molecules are not organisms. Hence, let us continue to look at the particles of living systems – but let us not forget life.

Bibliography

Ayala, F.J.: 'The Autonomy of Biology as a Natural Science', in A.D. Breck and W. Yourgrau (eds.), **Biology, History, and Natural Philosophy**, Plenum Press, New York, 1972, pp. 1-16.

Barker, E.: `Apes and Angels: Reductionism, Selection, and Emergence in the Study of Man', **Inquiry 19** (1976), 367-387.

Beckner, M.O.: 'Organismic Biology', in P. Edwards (ed.), **The Encyclopedia of Philosophy**, vol. 5, MacMillan, New York, 1967, pp. 549-551.

Bergson, H.: **L'evolution créatrice**, Alcan, Paris, 1907.

Bertalanffy, L. von: **General System Theory: Foundations, Development, Applications**, Braziller, New York, 1968 Penguin University Books, Harmondsworth, 1973.

Bunge, M.: 'General Systems and Holism', **General Systems** 22 (1977), 87-90.

Bunge, M.: 'Some Topical Problems in Biophilosophy', **J. Social Biol. Struct.** 2 (1979), 155-172.

Campbell, D.T.: 'Downward Causation in Hierarchically Organised Biological Systems', in F.J. Ayala and T. Dobzhansky (eds.), **Studies in the Philosophy of Biology** , MacMillan, London, 1974, pp. 179-186.

Cassirer, E.: **The Problem of Knowledge: Philosophy, Science and History Since Hegel**, Yale University Press, New Haven-London, 1950.

Cohen, J.: **Human Robots in Myth and Science**, George Allen & Unwin, London, 1966.

Cramer, F.: 'Fundamental Complexity: A Concept in Biological Sciences and Beyond', **Interdiscipl. Sci. Rev.** 4 (1979), 132-139.

Darwin, Ch.: **On the Origin of Species**, Murray, London, 1859 [The New American Library of World Literature, New York-Toronto, 1958].

Dobzhansky, T., Ayala, F.J., Stebbins, G.L., and Valentine, J.W.: **Evolution**, Freeman, San Francisco, 1977.

Driesch, H.: **Philosophie des Organischen**, Quelle & Meyer, Leipzig, 1928.

Eccles, J.C.: **The Human Mystery**, Springer International, New York-Heidelberg-Berlin, 1979.

Elsasser, W.: 'Principles of a New Biological Theory: A Summary', **J. theor. Biol. 89** (1981), 131-150.

Gould, S.J.: **The Panda's Thumb: More Reflections in Natural History**, W.W. Norton, New York-London, 1980.

Grene, M.: **The Understanding of Nature: Essays in the Philosophy of Biology**, D. Reidel, Dordrecht-Boston, 1974.

Grmek, M.: 'A Survey of Mechanical Interpretations of Life from Greek Atomists to the Followers of Descartes', In A.D. Breck and W. Yourgrau (eds.), **Biology, History, and Natural Philosphy**, Plenum Press, New York, 1972, pp. 181-195.

Himmelfarb, G.: **Darwin and the Darwinian Revolution**, W.W. Norton, New York, 1968.

Hull, D.: **Philosophy of Biological Science**, Prentice-Hall, Englewood Cliffs, 1974.

Jacobs, J.: 'Teleology and Reduction in Biology', **Biol. & Philos.** 1(1986), 389-399.

Kant, I.: **Kritik der Urteilskraft**, Hartknoch, Riga, 1790 [English translation: Hafner, New York, 1951].

Klerk, G.J.M. de: 'Mechanism and Vitalism: A History of the Controversy', **Acta Biotheor. 28** (1979), 1-10.

Kochanski, Z.: 'Kann die Biologie zur Physiko-Chemie reduziert werden?', in B. Kanitscheider (ed.), **Materie – Leben – Geist: Zum Problem der Reduktion der Wissenschaften**, Duncker & Humblot, Berlin, 1979, pp. 67-120.

Laszlo, E. (ed.): **The Relevance of General Systems Theory**, Braziller, New York, 1972.

Leinfellner, W.: 'Reductionism in Biology and the Social Sciences', in **Proceedings of the 13th International Conference on the Unity of the Sciences**, 1985 (in press).

Lenoir, T.: **The Strategy of Life: Teleology and Mechanics in Nineteenth Century German Biology**, D. Reidel, Dordrecht-Boston-London, 1982.

Lorenz, K.: **Evolution and Modification of Behavior**, The University of Chicago Press, Chicago-London, 1965.

Lorenz, K.: **Behind the Mirror**, Harcourt Brace Jovanovich, New York-London, 1977.

Marquand, J.: **Life: Its Nature, Origins and Distribution**, Oliver & Boyd, Edinburgh-London, 1968.

Mayr, E.: 'The Unity of the Genotype', **Biol. Zbl. 94** (1975), 377–388.

Mayr, E.: 'Teleological and Teleonomic: A New Analysis', **Boston Studies in the Phyilosophy of Science 14** (1974), 91–117.

Meyer-Abich, A.: **Ideen und Ideale in der biologischen Erkenntnis**, Barth, Leipzig, 1934.

Mohr, H.: 'Das Gesetz in der Biologie', **Freiburger Dies Universitatis 12** (1965), 23–49.

Mohr, H.: **Lectures on Structure and Significance of Science**, Springer, New York–Heidelberg–Berlin, 1977.

Polanyi, M.: 'Life's Irreducible Structure', **Science 160** (1968), 1308–1312.

Portmann, A.: **Neue Wege der Biologie**, Piper, Munich, 1965.

Portmann, A.: **Biologie und Geist**. Suhrkamp, Frankfurt/M. 1973.

Prigogine, I.: 'La thermodynamique de la vie', **La Recherche 3** (1972), 547–562.

Riedl, R.: 'A Systems-analytical Approach to Macro-evolutionary Phenomena', **The Quart. Rev. Biol. 52** (1977), 351–370.

Riedl, R.: **Order in Living Organisms: Systems-conditions of Evolution**, Wiley, New York, 1979.

Riedl, R.: 'A Dialectic Approach to Epigenetics and Macroevolution', in V.J.A. Novák and J. Mlikovsk'y (eds.), **Evolution and Environment**, CSAV. Prague, 1982, pp. 41–50.

Sattler, R.: **Biophilosophy: Analytic and Holistic Perspectives**, Springer, Berlin–Heidelberg–New York, 1986.

Schrödinger, E.: **What is Life?** Cambridge University Press, London–New York, 1967.

Simpson, G.G.: **This View of Life: The world of an Evolutionist**, Harcourt, Brace & World, New York, 1963.

Taylor, G.R.: **The Science of Life: a Pictorial History of Biology**, Thames &Hudson, London, 1967.

Vol'kenshtein, M.V.: **Molecules and Life: An Introduction to Molecular Biology**, Plenum Press, New York, 1970.

Wagner, G.P.: 'Feedback Selection and the Evolution of Modifiers', **Acta Biotheor. 30** (1981), 79–102.

Weiss, P.A.: 'Tierisches Verhalten als Systemreaktion', **Biologia Gen. 1** (1925) 168–248.

Weiss, P.A.: 'The Living System: Determinism Stratified'. **Studium Generale 22** (1969), 361–400.

Wilson, E.O., Eisner, T., Briggs, W.R., Dickerson, R.E., Metzenberg, R.L., O'Brian, R.D., Susman, M., and Boggs, W.E.: **Life on Earth**, Sinauer Ass., Stamford, Conn. 1974.

Wuketits, F.M.: **Wissenschaftstheoretische Probleme der modernen Biologie**, Duncker & Humblot, Berlin 1978.

Wuketits, F.M.: 'On the Notion of Teleology in Contemporary Life Sciences'. **Dialectica 34** (1980), 277–290.

Wuketits, F.M.: **Biologie und Kausalität**, Parey, Berlin–Hamburg, 1981.

Wuketits, F.M.: 'Systems Research –The Search for Isomorphism', in **Progress in Cybernetics and Systems Research**, vol. xi (1982). pp. 403–407.

Wuketits, F.M.: **Biologische Erkenntnis: Grundlagen und Probleme**, Fischer, Stuttgart, 1983.

Wuketits, F.M. (ed.): **Concepts and Approaches in Evolutionary Epistemology: Towards an Evolutionary Theory of Knowledge**, D. Reidel, Dordrecht–Boston–Lancaster, 1984.

Wuketits, F.M.: **Zustand und Bewußtsein: Leben als biophilosophische Synthese**, Hoffmann und Campe, Hamburg, 1985.

Wuketits, F.M.: 'Synthetic and Analytical Thinking', **Z. Anal. Chem. 326** (1987), 320–323.

Paul Hoyningen-Huene

EPISTEMOLOGICAL REDUCTIONISM IN BIOLOGY: INTUITIONS, EXPLICATIONS, AND OBJECTIONS

The term 'reductionism' is, no doubt, equivocal. Let me therefore first state what I shall address in this paper and what I shall not. Using the well-known distinction between ontological, methodological, and epistemological questions of reductionism (e.g. Ayala 1974; Hull 1981; Mayr 1982, pp.60-63), I will dismiss the **methodological** domain. Many of the questions of reductionism which are of interest to the working biologist **qua** working biologist, will therefore not be treated here. Furthermore, I will not address questions of the **ontological** domain: I shall stick dogmatically to ontological reductionism without arguing for it. For the sake of argument, I shall presuppose that living beings are composed of the same kind of matter that is equipped with the same kinds of elementary interactions that are known to physics and chemistry of today. Thus the questions of epistemological reductionism remain.

In this paper, I want to discuss some of the arguments that have been put forward, both for and against epistemological reductionsm. I shall do this in a slightly different manner from the usual one in which each party articulates its theses and presents arguments for it. A complex discussion exists in the literature which has led to points of disagreement that, at first sight, seem almost unrelated to the original topic and almost 'ideological' in character, despite the fact that the original topic seems perfectly clear-cut. My methodologically basic conjecture in this paper is that most of the dynamics of this discussion are understandable by turning not only to the thesis articulated by the respective party, but by additional reference to the underlying **intuition** for which the thesis is an **explication.** The **objections** to a thesis do not necessarily undermine the respective underlying intuition; they may rather stimulate a more sophisticated reformulation of that thesis. To the opposing party this may appear to be a stubborn, dogmatic, and consequently frustrating way of answering the objections. It is this interplay of intuitions, explications, and objections to which I now turn.

29

P. Hoyningen-Huene and F. M. Wuketits (eds.), Reductionism and Systems Theory in the Life Sciences, 29-44.
© *1989 by Kluwer Academic Publishers.*

I start with the position of the epistemological reductionist. His basic contention, as I understand it, is the following: Given the presupposition that living beings are made up of the same sorts of atoms and molecules as the physicists and chemists know them, and nothing else, then having a comprehensive knowledge of the particular atoms and molecules in a given organism, and their configuration, together with a knowledge of the laws of physics and chemistry, would in principle be sufficient to redefine all the properties of this organism in atomic and molecular terms, and to derive its behavior -- specifically, all the laws which it obeys. Generalizing this slightly, a knowledge of all the (relevant) properties of the elements of one level of the organizational hierarchy, together with a knowledge of how these elements are arranged at a particular higher level, would in principle be sufficient to redefine all the properties of, and derive the laws governing, the entities of this higher level. I will call this the **basic epistemological reductionistic intuition**, or more briefly, as we are in any case concerned only with epistemological questions of reductionsm, the `basic reductionistic intuition`. On the same rather intuitive level of expressing things, and asked why this would be really a plausible intuition, the epistemological reductionist would answer that given the knowledge of the components and their configuration, **everything** to be known about the system is known. Since no other factors which could be causally relevant to the behavior of the system can possibly exist, all knowledge about the system must, in principle, be extractable from the complete knowledge of the lower level. (For a decomposition of this basic reductionistic intuition into three parts see Oppenheim/Putnam 1958, pp.23-27.)

The basic reductionistic intuition leads to the straightforward and well-known logical empiricist explication of epistemological reductionism (see e.g. Nagel 1961): the higher level is epistemologically reduced to the lower level if and only if: 1. the higher-level descriptive terms can be, extensionally equivalently, expressed by lower-level descriptive terms (the `reduction functions`, `bridge principles`, or `correspondence rules`); 2. the higher-level laws can be logically deduced from the lower-level laws, including in the premises the reduction functions, and appropriate initial and boundary conditions describing the particular higher-level system (or class of systems) in lower-level terms.

Two remarks should be made here. Firstly, it is essential to include in the premises of this derivation initial and boundary conditions specifying the particular higher level system one is interested in. Omitting initial and boundary conditions would result in a completely uninteresting concept of reduction because it would be empty: no higher level, as a special configuration of lower level entities, could ever be epistemologically reduced to the lower level simply because the information of this special configuration would be missing, i.e. the system to be epistemologically reduced would not be specified. Although this point has been stressed many times (e.g. Nagel 1961, p.434; Putnam 1975, pp.138-139; Hull 1981, p.128; Friedmann 1982, p.38; Hoyningen-Huene 1985: but see also Hempel 1966 where initial and boundary conditions are not mentioned), an epistemologically anti-reductionistic position is sometimes argued for on the grounds that initial and boundary conditions are omitted from the premises of the deduction (see e.g. Ayala 1968, p.209; Ayala 2983, pp.284-285, 287; Dupré 1983).

Secondly, there has been some discussion about the status of the sentences that redefine higher-level terms, extensionally equivalently, in lower-level terms, the reduction functions. The problem consists of their status with respect to the analytic/synthetic distinction, with respect to the law/non-law distinction (e.g. Fodor 1974), and with respect to their alleged status as identities (e.g. Causey 1977). A minimalist position which seems appropriate here (but not in other contexts) simply leaves this question open: reduction functions have to be included in the premises in order to accomplish the deduction, whatever their status.

The basic reductionistic intuition together with its logical empiricist explication seemed for a while such a convincing line of thought that it was even used to explicate what ontological reductionism **meant** (e.g. Hempel 1966). In order to avoid ontological talk, one should **express** ontological reductionistic convictions in terms of epistemological reductionistic assumptions. This is highly plausible insofar as the way in which ontological convictions are gained in science, seems to consist of a comparison of the predictive and explanatory powers of theories with different ontological implications. The identification of ontological with epistemological reduction implied that it was impossible to articulate coherently a position that was both ontologically reductionistic (i.e. anti-vitalistic) **and** epistemologically anti-reductionistic.

But the logical empiricist explication of the basic reductionistic intuition soon came under attack, and with it the contention that ontological reductionism necessarily implies epistemological reductionism. There are a number of important objections to this account of epistemological reductionism, and I shall discuss some of them, especially with respect to their bearing on the basic reductionistic intuition.

1. With the advent mainly of Kuhn's and Feyerabend's work in the sixties it became clear that the logical empiricist explication was in trouble (Kuhn 1962; Feyerabend 1962). Firstly, even the simplest examples of accomplished epistemological reductions in physics demonstrate that the relation between reducing and reduced theory is not simply a logical deduction but involves

approximations (Schaffner 1967; also e.g. Wimsatt 1980, p.145). The reduced theory may, strictly speaking, logically contradict the reducing theory. Secondly, the same terms in the reduced and the reducing theory may have different meanings in the two theories, i.e. in the most obvious case, their extensions do not contain exactly the same elements.

It therefore became necessary to correct the explication of epistemological reduction: that is, the original, higher-level theory cannot be deduced from the lower-level theory together with the other premises, but only a `corrected` or `modified` version of it which is `strongly analogous` to the original theory (Schaffner 1967 and 1976).

How does this modification of the reduction model affect the underlying basic reductionistic intuition? It does not affect it at all; it shows only that reshuffling the information about the lower level, in order to yield knowledge of the higher level, is not as simple as the logical empiricists thought. The epistemological reductionist will not be intimidated by the prospect of rather complicated reduction functions in the epistemological reduction of classical genetics to molecular genetics, but he or she will not believe that reshuffling is impossible. Neither will he or she be intimidated by the relationship between molecular and classical genetics seeming more like a replacement than a reduction (whatever this difference is precisely), as long as it is believed that the ontological reduction holds (see especially Schaffner's 1976, pp.434-442 and Ruse's 1976, pp.447-454 reply to Hull).

2. A second objection concerns preconditions of scientific theories that have to be fulfilled in order to be able to decide whether the reduction relation between two theories holds or not.The theories in question must be sufficiently articulated to enable the (approximate) deductive relations to be proved or disproved. For this aim, theories have to be reconstructed by the philosopher, but this is not at all a unique procedure. Furthermore, since theories are historical entities they change continuously, and even at any particular point in time, many different versions of a theory may be held by different scientists. As a result, it is not even clear between which versions of theories a claim of epistemological reducibility holds, so that these claims are premature (e.g. Hull 1976 and 1981).

Again, all this may be conceded by the reductionist. Of course, earlier versions of a theory may not contain enough information about the lower level to accomplish the derivation: reduction functions may not be known. nor the initial and boundary conditions under which certain molecules perform certain function etc.. Also, it may be difficult to bring the theories to a degree of logical articulation that shows their connections explicitly. But all this does not damage the basic reductionistic intuition, it only involves more technical complications: if only atoms and molecules are involved, then the knowledge about them and their configuration is all that can be known about the system, however incomplete or insufficiently articulated for the logical connections, this knowledge and the knowledge of the upper level may be.

3. Further objections concern the status of the claim of epistemological reduction as an `in principle` claim. Nowadays no one would contend that an epistemological reduction of any field of biology has been worked out in such complete detail that it could simply be pointed at, and the discussion brought to a close. Rather, it is a claim that the epistemological reduction is `in principle` possible, given the ontological reduction.

Two main objections can be made. Firstly, the sense of `in principle` is not clear (see e.g. Wimsatt 1979 and 1980; Dupré 1983). Would a procedure that involved some 10^{120} discrete steps be counted as `in principle possible`? This is, of course, a good question. But again, it will not disturb the epistemological reductionist because there is no reason to believe, at least in the case of fairly simple biological systems, that such a large number of steps play a role in defining the reduction functions. (Things may be very different at the level of the human mind, but the problems of epistemological reduction of psychology should be distinguished from the problems of the epistemological reduction of biology). As long as such trans-astronomical numbers do not play a role, the fuzzy sense of `in principle` will not be disturbing.

Secondly, it has been stated that the epistemological status of `in principle` statements is problematic. They are believed to be "empirically meaningless" (Ayala 1967, p.210), or to be usually normative, as opposed to merely descriptive (Wimsatt 1979, p.358). But this does not really catch their epistemological character. I would suggest that the status of claims for the `in principle` possibility of some epistemological reduction should be regarded as similar in status to a **Gedankenexperiment**: though not directly empirically testable, it is not considered normative or even senseless. Rather its function is, roughly speaking, conceptual clarification by spelling out the ultimate consequences inherent in a certain approach (compare Kuhn 1977).

4. A further possible objection can be formulated as follows. Suppose that the epistemological reduction of biology (or some particular part thereof) to physics and chemistry was feasible. This would amount to the fact that all biological explanations would ultimately be physico-chemical. But this is highly implausible since biology is distinguished from the other natural sciences by its particular modes of explanation -- in particular, teleological explanation (e.g. Ayala 1968).

Therefore, the premise of this argument must be wrong, i.e. biology cannot be epistemologically reduced to physics and chemistry.

In order to handle this objection, we do not need to decide whether teleological explanations (or any other type of alleged specifically biological explanations) are truly the decisive factor for the autonomy of biology with respect to physics and chemistry. The above argument is wrong for a different reason, namely that it equates an epistemological reduction with an explanation (i.e. the validity of the Hempel-Oppenheim scheme is presupposed without qualifying appropriate conditions of adequacy). But certainly not every deduction from laws and initial and boundary conditions of a lower level, together with reduction functions, of a proposition about something at a higher level is an explanation of this something (see e.g. Putnam 1975; also Hoyningen-Huene 1985 for an analysis of the anti-reductionistic arguments in Popper 1974), and, conversely, the possibility of an explanation of specifically biological phenomena by lower level theories does not by itself imply the possibility of the epistemological reduction (Friedmann 1982). Thus, **explanations** of higher-level events and laws in terms of lower levels should not be identified with the **epistemological reduction** of the higher level to the lower level. Using the distinction of three different domains of problems of reductionism I used in the beginning of this paper, this is simply the difference between methodological and epistemological questions of reduction. Thus, even if biology was epistemologically reducible to physics and chemistry, this would not **ipso facto** constitute a threat to the autonomy of biology (compare Hull 1981, p.125; Friedmann 1982, p.37).

5. The next two objections are mainly put forward by biologists (e.g. Lorenz 1977, ch.3 [but see also ch.2 which does not seem to be consistent in all respects with ch.3, and Lorenz 1981, ch.1]; Mayr 1982, pp.63-64; Wilson 1975, p.7; also Popper 1974), and they rest on the concept of emergence (for a treatment of this concept in the recent philosophical literature see Pluhar 1978 and Klee 1984).

The first aspect of the idea of emergence as a counter-argument to epistemological reductionism is that the synthesis of some elements to some `whole` may lead to a behavior of that whole that is completely unexpected even on the basis of a complete knowledge of those elements, or their functioning in other combinations. As an example, Lorenz uses an electric circuit with a battery, a condenser, a coil, and a resistance connected in series. This circuit may exhibit damped oscillations which may indeed be unexpected if one knows the behavior of these elements only in connections of less than four elements.

But from this no argument against epistemological reduction follows. Though it is true that the oscillatory behavior of the circuit may be `unexpected` -- even in principle, it is not derivable from the knowledge of the components and their characteristics **alone** -- it is precisely derivable from their characteristics **and their connection**, i.e. the relevant boundary conditions. Thus, if this example is thought of as representative for an aspect of emergence, it is far from being a counter-argument to epistemological reductionism; on the contrary, it is an excellent confirmation for the basic reductionistic intuition (see also Hoyningen-Huene 1985).

6. But there is a subtler aspect of emergence that seems incompatible with epistemological reductionism which is also present in the above-mentioned writings of biologists. This second aspect appears at first sight to be almost identical with the first, but it can be developed in a different direction. The core of this aspect of emergence is that the elements of some `whole`may show a different behavior when integrated into that whole, from the behavior they display when examined separately. Thus the properties of these elements depend on the whole they are integrated into. and the knowledge of the whole is not derivable from knowledge of the elements studies in isolation. Conversely. if knowledge has to be gained about the elements, it must be derived from the knowledge of the whole, making epistemological reduction completely wrong-headed and impossible. This aspect of emergence stands somehow upon the same footing as the basic reductionistic intuition, and may therefore be called the **basic anti-reductionistic intuition**.

To this, the epistemological reductionist may, in order to save his intuition, respond with a distinction. namely the distinction between **gaining knowledge about** properties of things, and **the properties of these things themselves**. It may well be that the knowledge of the relevant properties of the elements of the whole, in order to accomplish the epistemological reduction of that whole. may be obtainable only by examining combinations of these elements. or even the very combination for which the epistemological reduction is intended. For example, the knowledge of a resistance relevant to its behavior in a circuit can only be attained by connecting it - really or ideally - with a battery. Similarly, the features of animals relevant to their social behavior can clearly be gained only by putting them into the appropriate social environment (or simulating it). Even so, the knowledge thus gained is knowledge about **properties of the elements** (appropriately called `dispositions`), even if it is obtainable only by investigating the whole, and by virtue of which an epistemological reduction may. however, be carried out legitimately.

Clearly, this argument appears to be self-immunizing to the anti-reductionist, and it may be parried by another distinction that has been put forward by epistemological reductionists themselves (Oppenheim/Putnam 1958. p.10: also Ayala 1983, p.286). This distinction is absolutely essential in order to distinguish true epistemological reductions from pseudo-reductions. The distinction is between predicates that truly belong to one level, and predicates that are **ad hoc** constructs in order to accomplish the epistemological reduction of the next highest level. The example of Oppenheim and Putnam for a predicate of the second sort is the predicate `tran` which should apply to **atoms** of transparent **substances**, and by means of which the transparency of macroscopic bodies can be reduced to the atomic level. (Obviously the example is faulty since the transparency of objects does not depend on the **sort** of constitutive atoms alone, but the idea of the distinction is clear). Now the dispositions introduced by the reductionist seem to be exactly of the second sort: they are only introduced to accomplish a (possibly even useless) epistemological reduction. Beyond that, their explanatory power is exactly the same as the "dormative potency" property of opium, as put forward by Molière's doctor.

But this move does not leave the reductionist defenceless; he or she may claim that the **criterion** by which the anti-reductionist classifies, for example, the social dispositions of animals as predicates of the second sort, as purely **ad hoc** in order to accomplish a reduction, is fundamentally mistaken. This can be shown by **reductio ad absurdum**, namely by using the same criterion for a class of predicates that are surely of the first sort, but which the criterion used by the anti-reductionist classifies as of the second sort. Take the case of any interaction of particles which is a fundamental situation for physics. Any sort of interaction between particles is explained in physics by the **charges** these particles bear, e.g. electromagnetic interaction by electrical charges of particles, gravitational interaction by (gravitational) mass of particles, etc. Charges as a fundamental property of particles are first introduced only to account for these interactions, that is a particle having a certain sort of charge, and being completely isolated from particles having the same sort of charge, may show no sign whatsoever of its charge without rendering the concept of its charge illegitimate. This is analogous to what biologists favoring emergence and hence epistemological anti-reductionism keep stressing: that their objects, e.g. some social animals, do not show certain properties when studied in isolation. But the analogy demonstrates that the introduction of certain predicates merely for the sake of the accomplishing a reduction is, by itself, insufficient reason to render these predicates purely **ad hoc** constructs, and thus illegitimate.

But why does the introduction of charges in physics seem legitimate even though they are at first introduced only in order to break down an apparently emergent property of an assembly of particles, and although these mysterious charges of a particle may never show up when the particle is separated from the respective whole? The answer is roughly this. First of all, these alleged properties of particles are very few, and, together with the laws of interaction, they account quantitatively for an infinite number of different wholes, i.e. they have a strong **predictive power**. Secondly, they obey certain conservation laws, i.e. they have a well-defined sort of **stability**. Thirdly, and largely as a consequence of the two foregoing items, they have **explanatory power**, i.e. it is believed that these charges are causally relevant and that means that they really exist.

Let us summarize the previous discussion about emergence and epistemological reduction. 1. The emergence of new properties of elements when integrated into some `whole` is, by itself, no argument against epistemological reductionism. 2. The attribution of properties to these elements that are potential in character when the elements are removed from the whole, for the sake of epistemological reduction, is not by itself unscientific even if it is in principle impossible to have evidence for the existence of these properties when the elements are removed from the whole. 3. The conditions for the legitimacy of the postulate of some properties that make epistemological reduction possible are, without the concrete case being specified, probably only vaguely determinable: the hypothesis of the existence of these properties should be credible, fruitful, independently testable etc., whatever that ma mean in a particular situation.

Thus the almost omnipresent fact of emergent properties as properties that show up only when the respective elements are integrated into some `whole` is, as such, completely neutral with respect to the epistemological reductionist`s and the anti-reductionist`s case. In order to argue their respective cases, the reductionist has to show that some emergent property is also potentially existent when the element is separated from the respective whole, and he or she has to do so, roughly speaking, by constructing an interesting and credible theory around the postulate of these dispositions. In the above-mentioned case, the disposition of members of a species to some social behavior, this might be the discovery of the genetic basis of this behavior, together with the discovery of the trigger mechanism which would make the lack of this behavior outside a specific environment intelligible, without making the **disposition** to it nonexistent.

The anti-reductionist must either show that the reductionist`s task is not realizable, or construct a theory which makes the emergence of the emergent properties intelligible **without** exclusive recourse to properties and laws of the lower level.

Does all this affect the basic reductionist`s intuition? Yes, it does so crucially because it makes explicit an assumption that went unnoticed as an integral part of that intuition, and that may be false. It is the assumption that indeed the behavior of an element of some `whole` can **always credibly** be traced back to some property this element also has in separation from this whole, if only as a disposition. Physics has been extremely successful with this, but this by itself is no argument that the success will continue forever.

7. The next objection can be formulated in a weaker and a stronger form. In the weaker form, it attacks the "tacit assumption of nomological determinism" that is asserted to be presupposed in the basic reductionistic intuition (Friedmann 1982, p.34). The thesis of nomological determinism claims that, given the set of laws governing the lower level, the set of laws governings the higher level is uniquely determined. However, in the case of the epistemological reduction of biology to physics and chemistry this thesis may not hold. Since one of the relevant theories is non-linear thermodynamics, it cannot be ruled out that the laws valid on the biological level may essentially depend on some microscopic fluctuation that took place during prebiotic or biotic evolution. This implies that the thesis of nomological determinism may be invalid in this case.

The obvious counter-move to this is to assert that the thesis of nomological determinism is not implied in the basic reductionistic intuition: the laws of the higher level are not simply determined by the laws of the lower level alone, but may depend on the relevant initial and boundary conditions as well. This counter-move may be parried in two ways.

The first possible parry is not very convincing: it contends that the above counter-move would deprive the biological laws of their character as laws, since their validity would then depend on some matter of fact (Friedmann 1982, p.24). But nobody I think would hold that biological laws as we know them, or are trying to know them, have to be completely independent of all events in the history of evolution. Take the `fact` (or `law`) of the chirality of biologically produced sugars, or some biological law depending on that fact. Nobody seems to believe that the chirality could not be opposite for life on a different planet, but this does not transform a generalization that is valid strictly for life on earth, and that depends on the chirality of biologically produced sugars, into a non-law. The concept of a biological law is simply not so strong.

The second possible parry seems cogent at first sight since it not only contends that the thesis of nomological determinism may be invalid, but also that even the laws of the lower level, plus the initial and boundary conditions, may be insufficient to determine the laws of the higher level (Friedmann 1982, p.35 fn.2 and p.38). This stronger thesis can be argued for on the basis of the indeterminacy of quantum mechanics. According to quantum mechanics, two identical systems, i.e. having exactly the same quantum mechanical initial state, may nevertheless develop into systems obeying **different** macroscopic, e.g.

biological. laws. A particular event, say the decay of a particle, may. for instance. take place at different times in the two systems leading to different microscopic fluctuations, which, in turn, may finally amplify to different macroscopic laws. (A variation of this argument not involving quantum mechanics may be obtained using identical **thermodynamic** initial conditions).

This presents a serious difficulty to the epistemological reductionist who may try to overcome it in the following way. but it is arguable whether this adjustment is justifiable, or purely **ad hoc** in a negative sense. In the above-mentioned case. the initial and boundary conditions **as they must be specified according to the laws of the lower level**. together with other necessary premises. are not strong enough to determine the (macroscopic) behavior of the system. and consequently. epistemological reduction is impossible. But this does not really run counter to the basic intuition itself. Microscopically speaking. if one cannot. in principle, predict what will happen, one cannot expect that. from all the available microscopic information (and the reduction functions). macroscopic behavior will be derivable **in every case**. This being the case. the missing microscopic information may be treated in the same way as other necessary factual information, namely the initial and boundary conditions in the technical sense. On the one hand, it may be argued that an unpredictable momentous microscopic fluctuation plays in its consequences the same role as the initial conditions in the technical sense, and must therefore be added · as supplementary information to the premises of the intended derivation. On the other hand. it may be argued that this move only proves the anti-reductionist's case. namely that from the physics and chemistry laws, together with all the information on initial and boundary conditions these sciences specify as such. and all necessary reduction functions. the biological laws will probably **in principle** not be derivable.

Which of these possible points of view one accepts as legitimate depends on the position one is defending. To the reductionist, the first argument seems fair. and the second unfair. although he will acknowledge the necessary correction of his or her position (unless some other way out is found). This is because indeterministic systems were not really covered by the basic reductionistic intuition, and some modification is therefore in order. To the anti-reductionist, it may seem quite the contrary, and the reductionist only adds one more self-immunizing move by which he or she is dogmatically trying to protect the position from devastating criticism. This situation is. in its circularity, very similar to situations which, according to Kuhn, occur in debates about paradigm choice: "Each group uses its own paradigm to argue in that's paradigm defense." (Kuhn 1970, p.94).

8. The last objection is different from the preceding ones in quality. It states that the whole approach to reductionism involving the concept of epistemological

reduction is at best uninformative, and at worst grossly misleading, about what is really going on in the life sciences when ties to physics and chemistry are made. As a consequence, philosophers of science should ask different questions, and should not remain stuck on a question that is in almost no contact with scientific life (see references in Ruse 1976, pp.454–460; Maull 1977; Wimsatt 1979 and 1980; Mayr 1982, p.63). This objection may be summarized in Lakatosian terms by saying that epistemological reductionism is the hard core of a strongly degenerating meta-scientific research program.

This is a very strong objection indeed to a philosophy of science that tries to analyze fairly heterogeneous reductionistic tendencies and research strategies in the life sciences just in the light of the one basic reductionistic intuition that leads to the doctrine of epistemological reductionism. Many other motives may underlie reductionistically inclined research, and because epistemological reduction and explanation are not coextensive, one may well mistakenly make epistemological reductionism the centre-piece of philosophy of biology.

How does this claim affect the basic reductionistic intuition? It does not, even if it is true, for two reasons. Firstly, one has to distinguish between this intuition itself, and the strategic place one alleges for it in the real process of research in the life sciences. For reasons that are easily understandable from the recent history of philosophy of science, epistemological reductionism may have played an unjustifiably pre-eminent role in the analysis of science (but see Schaffner 1981. p. 95). But removing it from this place does not imply that the underlying intuition itself is mistaken.

Secondly, it may be that it is worthwhile to keep an eye on the basic reductionistic intuition, and its possible role in the life sciences. Let me explain this by means of an analogy. Let us suppose that Kuhn is right and that something like `normal science` exists not just as a pitiable form of science. Does this falsify Popper's basic intuition that the spirit of science as a collective enterprise is, or at least should be, a critical one? I don't think it does, because all it demonstrates is that even in good science the idea of criticism is not at work as straightforwardly and directly as Popper thought. It may turn out that in order to really put into practise the idea of criticism, it may be necessary to refrain from it somehow, and that the idea of fundamental criticism cannot therefore be found in scientific everyday life, without making science basically uncritical. I am not suggesting that the situation is exactly analogous to the case of epistemological reductionism in the life sciences; I am only stressing the possibility that after a period of exaggerating the immediate importance of a certain idea, namely her the idea of epistemological reductionism, the danger of exaggerating the unimportance of it may be a real one in the period that follows.

So, what is the net result of this lengthy discussion? On the one hand, the basic epistemological reductionistic intuition has not survived the battle unscathed, and the most serious wounds come from quantum mechanics (objection 7), and the uncovering of the risky ontological conjecture implicit in it that properties of elements that are manifest only when these elements are integrated into some whole, may be treated as resulting from dispositions of these elements themselves (objection 6). Whether it bleeds to death from these wounds is difficult to say, but even if it does, it is not clear how many and which people will mourn, since its possible function as a Gedankenexperiment (objection 3), both in biology itself and the philosophy of biology, is not sufficiently clarified (objections 4 and 8). On the other hand, what I have called the basic anti-reductionistic intuition has come through the battle without a scratch only because it has been disqualified from it for not being relevant here (objections 5 and 6). And thirdly, both parties get into trouble when it comes to a precise articulation of what they defend or attack, respectively (objections 1 and 2), because each party may well survive the demolition of some explication of its respective intuition without being forced to surrender. Thus, if somebody pressed me to respond to the question: Who is right, the epistemological reductionist or the anti-reductionist? I would probably answer: yes and no.

Acknowledgement. Critical comments by Gerhard Vollmer on an earlier draft of this paper are gratefully acknowledged.

Bibliography

Asquith, P.D., Kyburg, H.E. (eds.): **Current Research in Philosophy of Science** (Edwards, Ann Arbor, 1979).

Asquith, P.d., Nickles, T. (eds.): **PSA 1982. Proceedings of the 1982 Biennial Meeting of the Philosophy of Science Association** (Philosophy of Science Association, East Lansing, 1983).

Ayala, F.J.: "Biology as an Autonomous Science", **American Scientist** 56 207-221 (1968).

--: "Introduction", in Ayala/Dobzhansky 1974, pp.VII-XVI.

--: "Beyond Darwinism? The Challenge of Macroevolution to the Synthetic Theory of Evolution", in Asquith/Nickles 1983, vol. 2 pp.275-291.

Ayala, F.J., Dobzhansky, T. (eds.): **Studies in the Philosophy of Biology. Reduction and Related Problems** (Macmillan, London, 1974).

Block, N. (ed.): **Readings in Philosophy of Psychology**, vol.1 (Harvard UP, Cambridge, 1980).

Causey, R.L.: **Unity of Science** (Reidel, Dordrecht, 1977).

Cohen, R.S., Hooker, C.A., Michalos, A.C., van Evra, J. (eds.): **PSA 1974** (Reidel, Dordrecht, 1976).

Dupré, J.: "The Disunity of Science", **Mind** 92 321-346 (1983).

Feyerabend, P.: "Explanation, Reduction, and Empiricism", **Minnesota** Studies in the **Philosophy of Science** III 28-97 (1962).

Friedmann, K.: "Is Intertheoretic Reduction Feasible?", **British Journal for the Philosophy of Science** 33 17-40 (1982).

Fodor, J.A.: "Special Sciences, or the Disunity of Science as a Working Hypothesis", in Block 1980, pp.120-133; originally in **synthese** 28 97-115 (1974).

Hempel, C.G.: **Philosophy of Natural Science** (Prentice Hall, Englewood Cliffs, 1966).

Hoyningen-Huene, P.: "Probleme des Reduktionismus der Biologie", **Philosophia Naturalis**, forthcoming.

Hull, D.: "Informal Aspects of Theory Reduction", in Sober 1984, pp.462–476; originally in Cohen et al. 1976, pp.653–670.

––: "Reduction and Genetics", The Journal of Medicine and Philosophy 6 125–143 (1981).

Klee, R.L.: "Micro-Determinism and Concepts of Emergence", Philosophy of Science 51 44–63 (1984).

Kuhn. T.S.: The Structure of Scientic Revolutions (Chicago UP, Chicago, 1962, 2nd. ed., enlarged, 1970).

––: "A Function for Thought Experiments", reprinted in Kuhn, T.S.: The Essential Tension. Selected Studies in Scientific Tradition and Change (Univ. of Chicago Press, Chicago. 1977), pp.240–265.

Lorenz, K.Z.: Behind the Mirror. A Search for a Natural History of Human Knowledge, transl. by R.Taylor (Methuen, London, 1977).

––: The Foundations of Ethology, transl. by K.Z.Lorenz and R.W.Kickert (Springer, New York, 1981).

Maull, N.: "Unifying Science without Reduction", in Sober 1984, pp.509–527; originally in Studies in History and Philosophy of Science 9 143–162 (1977).

Mayr, E.: The Growth of Biological Thought. Diversity, Evolution, and Inheritance (Belnap, Cambridge, 1982).

Nagel, E.: The Structure of Science. Problems in the Logic of Scientific Explanation (Harcourt, New York, 1961).

Nickles. T. (ed.): Scientific Discovery (Reidel. Boston, 1980).

Oppenheim, P., Putnam, H.: "Unity of Science as a Working Hypothesis, Minnesota Studies in the Philosophy of Science II 3–36 (1958).

Pluhar, E.B.: "Emergence and Reduction", Studies in History and Philosophy of Science 9 279–289 (1978).

Popper, K.R.: "Scientific Reduction and the Essential Incompleteness of All Science", in Ayala/Dobzhansky 1974. pp.259–284.

Putnam, H.: "Philosophy and Our Mental Life", in Block 1980. pp.134–143; originally in H. Putnam. Mind. Language, and Reality. Philosophical papers vol.2 (Cambridge UP. London. 1975) pp.291–303.

Ruse, M.: "Reduction in Genetics", in Sober 1984, pp.446–461; originally in Cohen et al. 1976, pp.653–670.

Schaffner, K.F.: "Approaches to Reduction", **Philosophy of Science 34** 137–147 (1967).

--: "Reduction in Biology: Prospects and Problems", in Sober 1984, pp.428–445; originally in Cohen et al. 1976, pp.613–632.

--: "Introduction [to] Reductionism and Holism in Medicine", **The Journal of Medicine and Philosophy 6** 93–100 (1981).

Sober, E. (ed.): **Conceptual Issues in Evolutionary Biology. An Anthology** (MIT Press, Cambridge, 1984).

Wilson, E.O.: **Sociobiology. The New Synthesis** (Belnap, Cambridge, 1975).

Wimsatt, W.C.: "Reduction and Reductionism", in Asquith/Kyburg 1979, pp.352–377.

--: "Reductionistic Research Strategies and Their Biases in the Unit of Selection Controversy", in Sober 1984, pp.142–183; originally in Nickles 1980, pp.213–259.

Michael Ruse

SOCIOBIOLOGY AND REDUCTIONISM

It is frequently said that sociobiology is reductionistic or, more precisely, since
much that is said about sociobiology is critical, it is frequently   said that
sociobiology is **unduly** reductionistic (Lewontin, 1977; Lewontin et al, 1984). In this
discussion, I want to explore the precise meaning of such a charge as this, and
whether or not it is justified. I hasten to add, at once, that I am not concerned
here with either defending or attacking sociobiology **per se**. I and others have had
much to say about this topic elsewhere -- some would say: "Too much to say
about this topic elsewhere." Rather, I want to do what perhaps should have been
done long ago.I want to provide a conceptual analysis of the way or ways in which
one might speak of sociobiology in the context of reductionism.

I shall begin by characterizing sociobiology very briefly. Then I shall move on to
an even briefer discussion of the various senses in which the term "reductionism"
is commonly used. After that, it will be possible to put the two concepts together
and to see what results.

SOCIOBIOLOGY

Sociobiology is the study of organisms, most particularly animals, in their social
relationships, from an evolutionary perspective (Wilson, 1975; Ruse, 1979).
Although the name is new, as a subject it goes back to the founding classic of
evolutionary theory, Charles Darwin's **On the Origin of Species** (1859). In that
work, Darwin argued that all organisms, living and dead, are the product of a
slow, gradual law-bound process of evolution, fuelled by a mechanism which he
labelled "natural selection". Darwin argued that many more organisms are born
than can possibly survive and reproduce, and that those which are successful in
life's struggle -- the "fitter" -- are, as it were, naturally picked out from the
breeding pool, for future generations. Given enough time, this natural selecting of

P. Hoyningen-Huene and F. M. Wuketits (eds.), Reductionism and Systems Theory in the Life Sciences, 45–83.
© 1989 by Kluwer Academic Publishers.

the successful leads to full-blown evolutionary change. What was crucial to
Darwin, and continues to be so for his followers, is that selection does not simply
lead to alteration, but promotes adaptation or proper functioning. Organisms work.
They have features and characteristics which aid their attempts to succeed in the
struggle for existence. They have eyes, fins, teeth, hair, feathers, leaves, roots,
branches -- all of these things are adaptations, helping their possessors in life's
struggles (Ruse, 1979b).

In the Origin, Darwin was quite explicit in his intention to apply his theory to
behaviour. What an organism does is just as much part of its nature as its
anatomy. But, Darwin wanted to go beyond mere behavior to social interactions,
particularly between animals of the same species. Darwin argued that, although life
is a struggle, this does not necessarily always imply hand-to-hand all-out conflict
-- "Nature red in tooth and claw", as the poet Tennyson had expressed it. Darwin
argued that cooperation between organisms can, under certain circumstances, be
just as much a product of evolutionary process as can any other characteristic.
Furthermore, Darwin backed up this belief by discussing in some length the
paradigmatic example of sociality in the organic world, namely the so-called social
insects (particularly the hymenoptera, the wasps, ants and bees). Darwin claimed
that the nature of the sterile workers in the hymenoptera must be just as much a
consequence of the selective process, as any other feature that we may encounter.

In the one hundred or more years since the Origin, evolutionary studies,
particularly neo-Darwinian studies, have exploded outwards and upwards with
great success. However, for a number of reasons, until recently the study of social
behavior by evolutionists lagged far behind. The chief reason, of course, was the
most obvious, namely that behavior is very, very much more difficult to study
than other aspects of organisms. It is much easier to tell something about the eye
colour of a fruit fly than about its mating habits. Thus, naturally, evolutionists
tended to concentrate on more readily discernible aspects of organisms, leaving
behavior (including social behavior) until later. Backing up this reason for the
neglect of behavior were strong ideological waves coming from the newly-growing
social sciences. For fairly straight-forward reasons, workers in such fields as
psychology and sociology felt very tense about the idea that biology might have
anything to contribute to their areas of inquiry. And so there grew up a strong
bias in favour of non-biological approaches to the understanding of social
behavior (even including, indeed especially including, studies of the social
behavior of non-human animals).

There were, nevertheless, some who looked at aspects of behavior from an
evolutionary perspective. Most notably, in the years before and after the last War,
the so-called "ethologists" looked at various questions to do with the interaction

between organisms. particularly social interactions between organisms of the same species (Lorenz. 1952: Tinbergen. 1953a, b). However, this was but a harbinger of what was to come, for some two decades ago (at the beginning of the 1960's) the whole study of the evolution of social behavior suddenly broke wide open. New ideas were introduced. new experimental results were gathered, new long–term naturalistic observations of animals in the wild were launched. From being the ugly duckling of the evolutionary family. the study of social behavior suddenly became a very proud and magnificent young swan indeed. In fact, it even gathered into itself a whole new name, namely "sociobiology". (A good introduction is Barash, 1982.)

INDIVIDUAL SELECTION

Sociobiologists are orthodox Darwinians. That is to say, in today's context, they believe that the major evolutionary motive force is natural selection, and that this works ultimately on the units of heredity. These units –– the "genes" –– are carried by organisms, within populations. and are transmitted through reproduction from one generation to another. All new variation within populations. on which natural selection must work ultimately, comes about through random changes ("mutations") caused at the level of the genes. There is. therefore, no overall direction or purpose to evolutionary change. What works, succeeds (Williams, 1966).

What distinguishes sociobiologists most strongly from their ethological and other predecessors (although, incidentally, not from Charles Darwin himself) is the extent to which they take an individualistic stance towards natural selection (Ruse, 1979a, 1980; Brandon and Burian, 1984). What does this mean? Consider for a moment the way in which natural selection might be presumed to act, and ask yourself the favourite question of lawyers, namely cui bono (who gains)? When a new adaptation is produced by selection. for whose benefit does the adaptation exist and function? Is the end worth of an adaptation directed towards the actual individual possessor of that adaptation, that is to say, the individual organism; or is the end worth of the adaptation directed towards the good of the organism's group, where "group", in most contexts, would be interpreted as the organism's interbreeding group (i.e., the organism's species)?

With some reservations and clarifications to be noted later, sociobiology's most distinctive characterizing feature is the extent to which it opts firmly for the individualistic solution (Dawkins, 1976). It is the claim of today's Darwinian students of social behavior that such behavior is caused by natural selection working at the level of the individual and not at the level of the group. Mutual aid, cooperation or, as sociobiologists like to call it, "altruism", must be understood as something which benefits the individual actors. It does not exist for the overall good of the species.

In taking such a stand as this, sociobiologists separate themselves firmly from ethologists, like the Nobel Prize winner Konrad Lorenz. In his well-known book, On **Aggression** (1966), Lorenz argued that the reason why animals never fight to kill is that such all-out combat would be harmful to the species. Sociobiologists challenge this interpretation, absolutely. They argue that any restrained aggression or like behavior must be understood in terms of individual benefits alone. In the particular case of restrained aggression, it is argued that although it might benefit the winner to kill the loser, all too often we ourselves are losers. Therefore, it is in our own interests that aggression be restrained (Maynard Smith, 1972, 1974).

Given that they take so individualistic a stance about selection, sociobiologists have found it necessary to devise various special models or mechanisms to account for particular facets of social behavior. It is here that they feel that they have scored their greatest triumphs. Amongst the various models which have been proposed we might, by way of illustration, mention the two which are most widely used. First, there is "kin selection". Here it is pointed out that the key to reproductive success lies in passing on one's own units of heredity, that is to say, in maintaining and perhaps increasing the proportion of one's own genes in future generations. However, obviously, one does not pass on one's own genes literally, but rather copies of such genes. But, remember that relatives will share genes. (By fairly elementary considerations one can show, for instance, that siblings are 50 % related, with corresponding diminution as the circle of relatedness widens.) What this all means, therefore, is that inasmuch as one's relatives reproduce, one is oneself reproducing vicariously!

If, for instance, my sister has children then, indirectly, she is poring my genes (or rather, copies of such genes) into the pool of genes in the next generation. Kin selection, therefore, suggests that there can be good biological reasons why one organism might aid the survival of reproduction of another organism. If the two organisms are related, inasmuch as the first organism is itself reproducing by proxy, as it were. Thus, we get the potential for the development of social behavior (Dawkins, 1976).

The second well-known proposed mechanism is so-called "reciprocal altruism". This is an idea which had its genesis even in the work of Darwin himself, and has recently been put to full use by sociobiologists (Trivers, 1971). Here there is no presumption that organisms are related. It is simply pointed out that, if there is reasonable expectation of such help in return, sometimes there are advantages in helping others. Thus, for instance, suppose we all stand a risk of drowning and that if no aid is forthcoming the chances of drowning are 100 %. Suppose, however, that if someone comes to our aid chances of drowning drop to 10 %, but that the price to be paid is the helper is itself also at a 10 % risk. It can be seen that there would be good biological reasons why an altruistic preparedness to aid others when drowning would be promoted by natural selection. The chances of oneself drowning drop from 100 % down to 20 % or so (given the way one specifies various aspects of the model). Note that here, as in the case of kin selection, there is no thought that social behavior directed towards others will evolve for the good of the species, or even for a more restricted group. The key to the species, or even for a more restricted group. The key to the evolution of such behavior is the way in which social attitudes and behavior will benefit the individual actor.

Perhaps the greatest triumph of sociobiology was also one of the earliest. William D. Hamilton (1964a, b) showed that the notion of kin selection can be applied to the hymenoptera, thus solving completely a problem which, as we have seen, was of concern even back at the time of Darwin. How does one get the evolution of sociality in the ants, the bees and the wasps? Why, for instance, should it always be the females which are the workers, rather than the males? Darwin (1859) felt certain that selection could not be benefiting species, but he felt that he had to allow that such selection could work at the level of the individual nest or hive.

Hamilton showed that one can take the analysis right down even to the level of the individual worker. The hymenoptera have a rather strange sexual system. Females have both mother and father, whereas males are produced asexually by females alone. This results in females having a full diploid chromosome set (half the chromosomes from the mother and half from the father), whereas males have only a haploid chromosome set (half the chromosomes from the mother). The consequence of this is that there are rather odd relationships within the hymenoptera family. In particular, females are more closely related to full sisters than to daughters. What this all means is that, from a biological point of view, females are better engaged in raising fertile sisters than fertile daughters. Thus, in evolving personal sterility, female worker hymenoptera are by no means dropping out of the reproductive race. What they are doing is promoting the maximization of their own genetic representation in future generations, in the more efficient way.

Thanks to their biology, it pays a female to raise a number of fertile sisters rather than an equal number of fertile daughters. Thus, through this application of kin selection, we can see how hymenopteran sociality evolved. There is, incidentally, no such biological virtue for males in raising siblings rather than daughters (males have no sons). Thus, we do not expect, nor do we find, the evolution of male sterility and a related preparedness to behave altruistically.

HUMAN SOCIOBIOLOGY

Let us turn now, equally briefly, to our own species, **Homo sapiens.** Charles Darwin always included us within the evolutionary spectrum. Indeed, in the first private speculations that we have been able to locate by Darwin on selection, the organism he used as an example was man. After the **Origin,** in the **Descent of Man** (1871), Darwin gave a full-blown exposition and analysis of human evolution, arguing that just as the social interactions of non-human organisms are brought about by natural selection, so likewise human social interactions are a function of the regular evolutionary process.

Sociobiology stand with Darwin on this and, although there are some variations, the best-known practitioners are adamant that natural selection must be brought to bear on all human social thought and behavior. Furthermore, this must be selection of an individualistic kind, not group selection of any sort. This means that, according to the sociobiological program, human care and kindness for others must be explained in terms of such causal models as kin selection, reciprocal altruism, and the like.

Perhaps the best-known of all the sociobiologists, the Harvard entomologist Edward O. Wilson, author of **Sociobiology: The New Synthesis** (1975), distinguishes between being what he refers to as "soft core" altruism, where one relates to others in expectation of returns, and "hard core" altruism where one relates to others but expects no such return. Wilson writes, in his work about our own species, On **Human Nature** as follows:

> Individual behavior, including seemingly altruistic acts bestowed on tribe and nation, are directed, sometimes very circuitously, toward the Darwinian advantage of the solitary human being and his closer relatives... Human altruism appears to be substantially hard-core when directed at closest relatives... The remainder of our altruism is essentially soft. The predicted result is a melange of ambience, deceit, and guilt that continuously troubles the individual mind. (Wilson, 1978, pp. 158–159).

Direct cooperation aside, in many other aspects the sociobiologists think that the ideas which were first developed in the context of the non-human world. For instance, if you take an individual selectionistic attitude towards sexuality, then obviously you do not see the relationships between the sexes, in particular between mates, as necessarily being one of untroubled harmony. Rather, one expects to see behavior and physical characteristics directed ultimately to the benefit of the individual, and not the pair. Thus, for instance (speaking now again just about the non-human world) one expects to find a certain amount of sexual dimorphism, because males and females are separate and one might expect them to have what have been termed "different reproductive strategies". Sociobiologists believe that this expectation of theirs is strongly born out, again and again, in the animal world. More often than not, males and females are physically, behaviorally, and emotionally different.

Analogously, the sociobiologists argue that the same holds of our own species. Quoting from Wilson's **On Human Nature**, once again:

> The anatomical difference between the two kinds of sex cell is often extreme. In particular, the human egg is eighty-five thousand times larger than the human sperm. The consequences of this gametic dimorphism ramify throughout the biology and psychology of human sex. The most important immediate result is that the female places a greater investment in each of her sex cells. A woman can expect to produce only about four hundred eggs in her lifetime. Of these a maximum of about twenty can be converted into healthy infants. The costs of bringing an infant to term and caring for it afterward are relatively enormous. In contrast, a man releases 100 million sperm with each ejaculation. Once he has achieved fertilization his purely physical commitment has ended. His genes will benefit equally with those of the female, but his investment will be far less than hers unless she can induce him to contribute to the care of the offspring. If a man were given total freedom to act, he could theoretically inseminate thousands of women in his lifetime.

> The resulting conflict of interest between the sexes is a property of not only human beings but also the majority of animal species. Males [in the majority of animal species] are characteristically aggressive, especially toward one another and most intensely during the breeding season. It pays males to be aggressive, hasty, fickle, and undiscriminating. In theory it is more profitable for ·females to be coy, to hold back until they can identify males with the best genes. In species that rear young, it is also important for the females to select males who are more likely to stay with them after insemination.

> Human beings obey this biological principle faithfully. (Wilson, 1978, pp. 124-5).

We hardly have yet a full exposition of sociobiology, nor yet of human sociobiology, but we have now sufficient of a "flavour" on which we can profitably base discussion. Let us, therefore, turn next to the other pole of our inquiry.

REDUCTIONISM

"Reductionism" is a term which is bandied about greatly by scientists, particularly
social scientists, and by the various commentators on science, including
philosophers, historians and sociologists. Of course, as so often happens when a
term is much used, different people tend to mean different things and, unless
great care is taken, one spends much time unprofitably talking at cross-purposes.
For the sake of clarity in this discussion, I shall imply an enlightening taxonomic
division by the well-known population geneticist, Francisco J. Ayala (1974). He
has argued, with some insight, that one can properly subdivide the notion of
reduction into three rather separate senses.

First, we have what Ayala speaks of as **ontological** reduction. He writes as follows:

> From the **ontological** point of view, the question of
> reduction is whether physico-chemical entities and
> processes underlie all living phenomena. Substantive
> vitalists claimed that living processes are, at least part,
> the effect of a non-material principle or entity, which was
> variously called `vital force`, `entelechy`, **`élan vital`**,
> `soul`, `radical energy`, or the like ... Ontological
> reductionism implies that the laws of physics and
> chemistry fully apply to all biological processes at the
> level of atoms and molecules" (p. viii).

Next, Ayala speaks of what he terms **methodological** reductionism. Here we are
speaking more of an attitude, or strategy, that the scientist employs. He writes as
follows:

> What I have called ... the **methodological** domain
> encompasses questions concerning the strategy of research
> or the acquisition of knowledge. In the study of life
> phenomena, should we always seek explanations by
> investigating the underlying processes at lower levels of
> complexity, and ultimately at the level of atoms and
> molecules? Or most we seek understanding from the study
> of higher as well as lower levels of organization? (p. viii).

The third kind of reduction Ayala speaks of as **epistemological** reductionism. This,
as he correctly notes, is the type which has been most fully discussed in the
philosophical literature, over the years. Once again quoting Ayala:

> **Epistemologically**, the general question of reduction is
> whether the theory and experimental laws formulated in
> one field of science an be shown to be special cases of
> theories and laws formulated in some other branch of
> science. If such is the case, the former branch of science
> is said to have been reduced to the latter (p. ix).

To be honest, I am not sure that these various senses of reduction are all unambiguously separate (nor, I suspect, did Ayala intend to convey that impression). No doubt, someone who was a methodological reductionist would (for instance) feel very uncomfortable with denial of ontological reductionism. However, Ayala does surely correctly capture some of the various notions which enter into discussions about reductionism, and certainly his division will prove useful to us as we now turn to discuss the main problem being posed in this essay. In what sense or senses can we say that sociobiology, including human sociobiology, is reductionistic? More precisely, in what sense or senses can we say that sociobiology, including human sociobiology, is unduly reductionistic? Let us structure our analysis by taking the sociobiology sketched in the last sections and comparing it against the various notions of reductionism, as just given in this section.

SOCIOBIOLOGY AND ONTOLOGICAL REDUCTION

If one is a sociobiologist, is one committed to some form of ontological reductionism? In other words, if one is a sociobiologist does one believe that, ultimately, everything comes down to atoms or molecules, or some such thing? Or, does one rather thing that there are other forces or entities of an ethereal life-form, operative in organisms, which must be understood before one can have a full understanding of their nature?

As far as the non-human world is concerned, at least as far as that large part of the non-human world not closely related to us humans is concerned, there seems little doubt that ontological reductionism is assumed quite happily by all. Like virtually any other biologist working today, no sociobiologist is going to entertain at all the hypothesis that, for instance, the hymenoptera in their various social interactions are in any way guided by mysterious life forces or such things. In the opinion of Wilson and his fellow students, when one gets down to the basic levels, ants, bees and wasps are no more than molecules moving around in space. There is no "élan vital" making workers help their sisters in the nest, or any such thing.

The precise status of such an ontological reductionistic stance is, I suppose, a matter of some debate. No doubt most biologists would think that the notion that vital forces might be involved is simply false. I expect that a neo-vitalist, anti-ontological reductionist would protest that the actions of such forces are not such that could be detected by ordinary materialist means. In which case, the response of sociobiologists would be that of most other people (including most philosophers today), namely that even if vital forces exist they seem to be totally unnecessary in an understanding of organic nature (Hempel, 1966). Therefore, they are at the very least totally redundant. Thus, ontological reductionism is justified on grounds of simplicity, if nothing else. But, however one makes the final decision, descriptively, it is indubitably the case that sociobiologists (when dealing with non-humans) are ontological reductionists. Relatedly, prescriptively, sociobiologists feel no reason to change their mind in this. Nor, one might add, does one see any desire in the attacks of critics of sociobiology to combat this kind of reductionistic stance.

What about our own species (and, perhaps, what about organisms close to us, like the higher apes)? There are undoubtedly those today who, although not old-fashioned vitalists, would oppose a full-blooded ontological reductionistic attitude towards human beings. I think here particularly of those such as John Eccles and Karl Popper (1977), who take a "dualistic" attitude towards human consciousness. In the tradition of René Descartes, these thinkers argue that in order to understand the human mind, one must suppose that there exists a kind of substance which is not of, nor reducible to, material substance. For such people as these, therefore, the thesis that humans are no more than molecules down at the substantival level, is simply inadequate. For a full understanding of humans, and include here the higher apes, if you will, one must suppose more than the purely material.

This is very much a minority position today. Most thinkers about human nature, whether or not they be sympathetic to human sociobiology, tend to feel that the supposition of some sort of Cartesian immaterial thinking substance raises more problems than it solves (Churchland, 1984). Most would argue that the hypothesis that consciousness is something other than, and quite distinct from, the material leads to no true insight -- certainly not at the scientific level -- and only creates problems about how mind and body are supposed to interact. Thus, working scientists of all persuasions generally take an ontological reductionistic stance, even towards human beings. They opt away from Cartesian dualism, and generally in favour of some sort of psycho-physical identity thesis, where mind and body are thought of as being, in some respects, different aspects of the same thing. (Of course, there are many different suggestions and metaphors which are used today -- one very popular one is to think of the brain as the kind of hardware of thinking, and the mind as the software. I need take no stand on this kind of suggestion here.)

Where stand sociobiologists on the issue of ontological reductionism, especially qua our own (and perhaps closely related) species? It is true that sociobiologists have not always shown the full sensitivity to the problems of the mind that one might have hoped (and certainly find in the work of other writers). Nevertheless, to the best of my knowledge no sociobiologist writing about our own species has ever strayed far from the belief that the human mind is ultimately a function of material objects like molecules. In other words, like most other students of humankind, sociobiologists take ontological reductionism, even with respect to human beings, as a given and then work out from there.

Thus, for instance, recently Wilson, in collaboration with the young physicist Charles Lumsden, has developed a fairly complex theory of human nature which supposes that our thinking is constrained by certain innate bounds (Lumsden and Wilson, 1981, 1983). These constraints Wilson refers to as "epigenetic rules", arguing that they have a physical, material basis in the brain and are put in place, guiding and molding our thinking and behavior, for their adaptive value. Illustrating his position, Wilson argues that we have epigenetic rules affecting our senses, for instance those of taste. Humans have an innate liking for certain things, like honey, and a distaste for sour and decaying would-be foodstuffs. There are obvious biological reasons why we should thus respond to possible sources of nutrition. Those proto-humans who preferred ripe apples over sour apples tended to survive and reproduce more efficiently than those that did not.

Analogously, at a more abstract level, Wilson and Lumsden argue that we have certain predispositions to avoid incestuous relationships with our siblings. The reason for this disinclination lies in the horrendous biological effects that follow on close inbreeding. Brother-sister offspring tend to be really sick human beings. What Wilson argues, therefore, is that selection has made us in such a way that we simply do not want to go to bed with our siblings (in contrast to our behavior and attitudes towards other members of the human species where, given the high premium on reproduction, sexual inclinations work at high level).

It is clear that this theory of epigenetic rules, which is only just at the beginning point of being worked out, takes a firmly reductionistic stance from the ontological perspective. Whatever else may be said, neither Wilson nor Lumsden have any doubt but that ultimately we are talking about molecules as they come together in cells in the human body, and about the consequent effects of such material objects. There is no supposition that the epigenetic rules in any way transcend or violate or stand outside the laws of physics and chemistry. Of course, as we shall see later, this is not to say that humans are no more than molecules, in every sense that one might ask the question about them. However, it is to say that, from the sociobiological perspective, the assumption is that, in dealing with humans, we are dealing with natural, physical objects, no less than we were dealing with pendulums, prisms, chairs, tables, rocks or planets.

Hence, we see that, from the viewpoint of ontological reductionism, sociobiology, including human sociobiology, makes a firm commitment to the ultimate reality only of material particles. Biologists themselves would probably think this is an empirical assumption. It is perhaps something rather stronger than this -- an assumed precondition for thinking biologically about organisms, including ourselves -- but whatever the ultimate status one accords to ontological reductionism, the sociobiological attitude towards it seems in no way exceptionable.

In short, completing the first stage of our analysis, we can agree that, in the respects being discussed, sociobiology (including human sociobiology) is highly reductionistic. However, there is little reason, as such, to condemn it for being **unduly** reductionistic.

METHODOLOGICAL REDUCTIONISM

Turn now to questions about levels of organization and appropriate strategies for explanation. Turn now to questions of methodological reductionism. Do sociobiologists have an inclination or tendency to explain the more complex in terms of the more simple, the group in terms of the individual, the larger in terms of the smaller? If they do, how far do they want to take this strategy? Do they want to go right down to molecules, as the ultimate goal in their efforts to provide explanations of social behavior? Is human social behavior to be explained in terms of quantum mechanics, for instance? Depending on the answers to these various questions, are the sociobiologists right or at least justified in the attitudes and directions they take?

It is here, of course, that we start to get to the real reason why the charge of reductionism is so often raised in discussions about sociobiology. Although sociobiologists are ontological reductionists, as we have seen there is nothing particularly exceptionable in their taking this attitude. Indeed, the oddness would be if they thought otherwise. However, methodological reductionism is more controversial. No one would deny that, as a strategy in science, it has lead to great results. Imagine where molecular biology would be had there been no such methodological reductionism. Even those who were hitherto quite hostile, now appreciate that such a strategy has paid great dividends in our understanding of the principles of heredity (Mayr, 1982). Going from old Mendelian genetics to new molecular genetics has taught us much about organisms -- much which was hitherto concealed from us.

However, whether methodological reductionism is always a good thing is, in the opinion of many, an entirely different matter. They would argue that perhaps the obsession with the small and the simple and the decomposed conceals from one important aspects of reality -- aspects which are apparent only at a global or more complex or more integrated or holistic level (Lewontin, 1974). I shall not get into general discussion, saying only that, clearly, defenders of methodological reductionistic strategies are right in saying that such approaches frequently led to major advances in science; and that perhaps also opponents of methodological reductionism are right in saying that it is not logically necessarily the case that methodological reductionism is the uniquely right strategy. Perhaps there are higher levels of understanding which require a more holistic approach. At least, this is my a priori intuition, which I shall now bring to bear in looking at sociobiology.

One point needs no argument at all. Above all else, what is distinctive about sociobiology (I speak now of both non-human and human varieties) is a strong commitment to methodological reductionistic strategies. As we have seen above, it is such an attitude towards the understanding of social behavior that separates out sociobiologists most sharply from other biological students of social behavior, like the ethologists. The sociobiologists want to explain at the level of individual selection, not at the level of group selection. Sociobiologists want biological advantage to be reduced down to the single organism against the world, as it were. Any higher levels of sociality, for instance, as one finds in the hymenoptera, must in some way be made understandable and reducible down to motives or actions or strategies at lower levels. There is absolutely no place for wholesale group selection hypotheses. And, this is obviously methodological reductionism, brought to a high pitch.

Now, granting this strong commitment to this type of reductionism, something which we have seen to be as characteristic of human sociobiology as of non-human sociobiology, the next questions concern the status of such reductionism and, consequent upon this, whether or not such a reductionistic strategy is a good or bad thing. Take first the question of status. Certainly, the example of the ethologist shows clearly that one does not necessarily have to be such a reductionist in tackling questions of social behavior from a biological perspective. Indeed, even back at the time of Darwin, there was debate about whether or not the individual selectionistic approach, which is so characteristically methodologically reductionistic, is the essential or most fruitful approach in evolutionary understanding. Natural selection's co-discoverer, Alfred Russel Wallace, differed sharply from Darwin over this question, arguing that higher levels of selection are, at times, both needed and appropriate (Ruse, 1980).

The question we must ask, therefore, since apparently choice is, or at least was, open, is: Why do sociobiologists so strongly favour the methodological reductionistic approach? Or, since, by definition, sociobiologists seem to be those who do favour such an approach, perhaps one can more precisely frame the question as: Why do those people who have come to be associated with sociobiology take so strong a methodological reductionistic approach?

If one turns to the literature, looking for times when sociobiologists talk self-consciously about their approach, one usually finds that they justify their attitudes on the grounds of elegance or simplicity or some such non-cognitive basis (Williams, 1966). It is argued that one ought, as scientists -- at least as good scientists -- always prefer the explanation in terms of lower levels of organization, over the explanation in terms of higher levels of organization. The choice must be for simple over complex, for the part over the whole. And, in justification of this scientific exhortation, something akin to Occam's Razor is usually invoked. The appeal to simplicity is taken to be justification itself (Brandon and Burian, 1984).

There are, I suspect, a number of points which could be raised in query here. For instance, one might object that in lumping together such notions as higher and lower levels of organization, simplicity and complexity, and so forth, one is in fact conflating a number of rather different things. For instance, one can well imagine a more holistic approach significantly simpler than an approach which appealed to parts rather than wholes. A good example is the way in which Boyle's Law makes easy macroscopic sense out of a lot of very, very complex interactions occurring at the microscopic level.

Furthermore, as critics point out, although the preference for simplicity is frequently a good strategy, it is obviously more a function of our own limitations than a reflection of the way that the world must be. There is certainly no guarantee that the world must be simpler rather than more complex. Specifically, in the case of social behavior, there is no a priori reason why individual selection must necessarily always take preference over group selection.

However, fortunately at this point, as so often happens, we can rescue the empirical scientists from their own philosophical commentaries on their own work! In fact, if one looks at the actual practise of sociobiologists, one finds that the real reason for the individualistic stance is rather stronger than a straight appeal to Occam's Razor, or some such thing. The fact of the matter is there are very good theoretical arguments, backed now by strong empirical evidence, suggesting that other than under rather exceptional circumstances group selection simply will not function efficiently.

The reason was realized by Charles Darwin himself, and has been elaborated and confirmed again and again by sociobiologists (Dawkins, 1976). The organism with an adaptation of benefit to the group, but which adaptation in some sense counts against itself, is going to be at a selective disadvantage to an organism with an adaptation of benefit to itself, and not at all or only incidentally to the group. The second organism will outreproduce the first organism. Consequently, in the next generation only the second organism will be represented, and however much good the first organism's group-benefiting characteristics may have been, they are gone and lost forever.

The trouble with natural selection is that it is entirely short-sighted. It does really not matter at all how much good something will do down the road, if the immediate pay-off is not strong enough. Put the case this way: Suppose organism "A" has a characteristic which gives the group a general pay-off of 10 % (per member) but which aid is done at a cost of, say, 50 % to itself; and suppose the second organism "B" has a characteristic which does itself 20 % good and nothing at all for the group, or perhaps even a negative pay-off for the group. Organism "A" will simply be outreproduced by organism "B". Indeed, "B" will take advantage of "A"'s help, without even having to reciprocate! That this theoretical consideration holds empirically has been demonstrated again and again both experimentally and empirically by sociobiologists. Note that even the group selectionist must eventually translate group benefits into individual benefits. To talk of a group benefit which aids no individual is meaningless.

We can see, therefore, that there are general theoretical considerations, suggesting that individual selection will (much more usually) work rather than group selection. Moreover, today, there is much empirical evidence (drawn from nature and from experiments) backing these considerations. It is for these reasons that the sociobiologists prefer the individual selectionist approach. In other words, they are not methodological reductionists because they appeal to some vague metaphysical or esthetic principles governing science. Rather, they take such a strategy for the most straightforward reasons of whether or not their theories and models accurately reflect what appears to be the case in the real world.

Of course, you might point out that, having truly described cases where individual selection works, sociobiologists then go on to assume that it works fairly universally. Furthermore, even when the reasons for an individual selection mechanism are not apparent, sociobiologists assume that such a mechanism holds nevertheless. In other words, when faced with unexamined instances of social behaviour, the presumption of sociobiologists is that an individual selectionist explanation is appropriate. This is presupposed by sociobiologists even before they start study. And, this being so, you might object that, for all the apparent fair-mindedness, sociobiologists do go illicitly beyond their evidence. Although perhaps they have some justification for methodological reductionism in those cases which they have explained. sociobiologists show hidden metaphysical or ideologically commitments when they turn to the unknown. In other words, looking at the overall perspective of sociobiologists, there still seems to be an innate, unjustified bias towards methodological reductionism.

That sociobiologists do have now have such a preference for individual selectionistic explanations, I fully admit. (I use the word "preference" rather than "bias" or "prejudice", as some critics do. Otherwise, there is an unfair loading of the case against sociobiologists.) However, in noting this preference, I would deny absolutely that it is unjustified, or "metaphysical" in a perjorative sense. If one has found that a particular strategy pays rich dividends in enabling one accurately to understand and model the world. then it is just plain foolish to turn one's back on it as soon as one presses out into the unknown (Ruse, 1979a; 1982).

Indeed, as philosophers and others are now realizing fully, success in science demands that one goes beyond the known to the unknown, and that the way in which one does this is by pushing one's already-established theories into unexplored domains (Kuhn. 1962). This is what Newtonians did so successfully. This is what modern physicists do so successfully. This is what other scientists, like geologists, do all of the time. Only by invoking altogether different and more stringent criteria for the sociobiologists can one deny them the right to do precisely the same, or fail to praise them when they do this. Individual selection approaches have worked most magnificently, as Hamilton's work on the hymenoptera proves fully. Therefore, it is sound scientific practise to push such a strategy as far and as hard as one can.

Backing this point is the fact that sociobiologists show, through their work and attitudes, that they are not dogmatically committed to methodological reductionism. In fact, there is recognition that under certain, undoubtedly special circumstances, group selection might occasionally function effectively (Wilson, 1975). For instance, if you had very small, fragmented populations, and organisms carrying generally benevolent or altruistic features might possibly affect the good of their groups before individual selfishness takes over. Then, the group-aiding features could spread right through the species, as the unsuccessful groups died out. Whether such situations occur very commonly is a highly debatable point; but, the simple fact of the matter is that sociobiologists are aware of the theoretical possibility that occasionally such group effects might take place. Furthermore, as genuine, vigourous scientists, sociobiologists explore these notions both in theory and in experiment. Thus, although individual selectionism is by far the norm amongst sociobiologists, their willingness to toy with selection at higher levels suggests that there is nothing unduly dogmatic about the commitment to methodological reductionism. (See Brandon and Burian, 1984, for more details.)

GENIC SELECTION

We cannot stop our discussion here. We must go on to another, perhaps-even-deeper aspect of the question. I have spoken of sociobiologists being committed to a policy of methodological reductionism, equating this with a commitment to the general efficacy of individual selection. But, what is meant by "individual" in this context? Thus far, I have assumed (and, indeed, asserted) that individuals are organisms. However, as one looks into the sociobiological literature, it is clear that many sociobiologists themselves want to take their argument one step further. They speak in terms of "genic selection", meaning that ultimately the individual is not the living organism but rather the units of heredity themselves. That is to say, the individuals are Mendelian genes or perhaps strips of D.N.A.

Here, obviously, the sociobiologists take their methodological reductionism to a further pitch than I have hitherto indicated. Thus, for instance, we find Richard Dawkins (1976), speaking of "selfish genes", writing that "a body is the genes' way of preserving genes unaltered". He adds, in more detail:

> We are survival machines -- robot vehicles blindly programmed to preserve selfish molecules known as genes
> ...
>
> They swarm in huge colonies, safe inside gigantic lumbering robots...
>
> They are in you and me; they created us, body and mind; and their preservation is the ultimate rationale for our existence. (Dawkins, 1976, p. 21, quoted by Gould, 1980, p. 122).

Making absolutely clear just how firmly he is committed to an extremely strong methodological reductionistic approach, Dawkins clarifies as follows: "I shall argue that the fundamental unit of selection, and therefore of self-interest, is not the species, nor the group, nor even, strictly, the individual. It is the gene, the unit of heredity" (Dawkins, 1976, p. 12, quoted by Gould, 1980, p. 122).

What is objectionable about this? If the organism is a more appropriate unit of selection than the group, why should not the gene be a more appropriate unit of selection than the organism? Critics respond that matters are not quite analogous. Organisms are separate entities which can, as it were, act independently of their fellows. For instance, if I have ten offspring, I might make you pleased or unhappy. But, my having ten offspring does not make much difference to your fertility -- at least, not in any very direct way. Neither would your fitness be much affected by my committing suicide. However, genes occur in far more integrated situations. Genes are not separate from each other. Rather, they occur organized on the chromosomes, within the cells of organisms. What happens to one gene is usually - virtually inevitably -- very much a function of what happens to other genes. Usually, the potential that one gene has for survival and reproduction (that is to say, for transmission to the next generation) is not independent of the potential that its fellow genes have. As Ernst Mayr (1975), a biologist with impeccable Darwinian credentials, has long argued, one must think of the collection of genes within the cell -- the genotype -- as an integrated unit.

Hence, the critics argue, one simply cannot speak meaningfully of genic selection, because this implies the genes are separate individuals. In reality, genes occur always as parts of integrated wholes. Consequently, selection can only work on the whole, not the particular. Against Dawkins, Steven J. Gould writes as follows:

> No matter how much power Dawkins wishes to assign to genes, there is one thing that he cannot give them -- direct visibility to natural selection. Selection simply cannot see genes and pick among them directly. It must use bodies as an intermediary. A gene is a bit of DNA hidden within a cell. Selection views bodies. It favours some bodies because they are stronger, better insulated, earlier in their sexual maturation, fiercer in combat, or more beautiful to behold.
>
> If, in favoring a stronger body, selection acted directly upon a gene for strength, then Dawkins might be vindicated. If bodies were unambiguous maps of their genes, then battling bits of DNA would display their colours externally and selection might act upon them directly. But bodies are no such thing.
>
> There is no gene "for" such unambiguous bits of morphology as your left kneecap or your fingernail. Bodies cannot be atomized into parts, each constructed by an individual gene. Hundreds of genes contribute to the building of most body parts and their action is channelled through a kaleidoscopic series of environmental influences: embryonic and postnatal, internal and external. Parts are not translated genes, and selection doesn't even work directly on parts. It accepts or rejects entire organisms because suites of parts, interacting in complex ways, confer advantages. The image of individual genes, plotting the course of their own survival, bears little relationship to developmental genetics as we understand it. (Gould, 1980, p. 123).

Consequently, conclude the critics, the methodological individualism of the sociobiologists does them a disservice. Such students of social behavior can argue for the ultimate kind of selection that they favour, only by ignoring vital facets of the organic world. Indeed, they make their case only by ignoring some of the most elementary principles of genetics and developmental biology.

What can we say about this criticism? Is it as effective against the very foundations of the whole sociobiological programme, as many critics rather imply? My feeling is that, if nothing else, the critique shows many sociobiologists can rightly be criticized for sloppy thinking and language. There is absolutely no doubt that genes occur in integrated situations, which are far more confining than the situations in which organisms usually occur. More often than not, it makes little sense to talk about the effect of some particular gene. What one must look at is the gene considered against the whole collection of genes within an organism's cells. However, I am far from convinced that one has here an absolutely devastating critique of sociobiology -- continuing to speak now of both human an non-human varieties. One cannot yet argue that the sociobiologist's methodological reductionistic attitudes simply cause everything to collapse in a cloud of confusions and contradictions.

For a start, there is empirical evidence suggesting that, in its most reductionistic interpretation, genic selection may not be without its merits. It is true that by the time the genes show their effects on the physical characteristics of organisms (the "phenotypes"), these effects usually vary according to the other genes in the organism. Nevertheless, there is a growing suspicion that, beneath the phenotype, there is competition between the genes, leading to a kind of genic selection. Recently, a number of molecular biologists have argued that much D.N.A. shows little or no effect on the actual physical characteristics of the organism. Suppose one has two strips or such D.N.A., with one strip possessing a much greater propensity to increase in size and proportion than the other. Then, obviously, what one gets is a kind of natural selection down at the level of the D.N.A., or the genes. This is clearly "genic selection" of some kind, with success going to the most "selfish DNA". Furthermore, it is one which molecular biologists tell us is increasingly likely (Doolittle and Sapienza, 1980). So it would be dangerous, at this point, to insist with the critics that methodological reductionism has misled us completely.

However, in fairness, it must be agreed that this kind of genic selectionism was hardly what sociobiological defenders like Dawkins had in mind. What of their variety, which does have phenotypic implications, and which was at the focus of the critics attack? Here I think it is appropriate to draw a distinction, as Dawkins himself has drawn, between the genes and the situation within which they find themselves. Dawkins, with his happy gift for metaphor, speaks of the distinction between "replicators" and "vehicles" (Dawkins, 1978). Genes are replicators. It is they that get copied over and over again down through the generations; they which, as it were, constitute the evolutionary thread from one to the next. And it is, as orthodox Darwinian theory would have it, genes (or rather the rations of genes) which constitute evolutionary change. If you have a change from one genetic form to another over a number of generations, then you have evolution. Individual organisms are vehicles. It is they which carry genes about, and which help the genes to transfer from one generation to another. Naked genes, sitting on toadstools, will get nowhere. Genes need to be packaged together within organisms and, thus, they have an opportunity to transmit themselves from one organism or vehicle to another.

Now, as Dawkins points out, natural selection generally acts at the level of vehicles. In this sense, natural selection is organism-level individualistic. However, the effect of natural selection is not merely on vehicles -- or even, from an evolutionary perspective, most essentially on vehicles. Rather, selection's effect is on replicators. Evolutionary change requires change in replicators. In this sense, natural selection is genic-level individualistic, not with standing the fact that such selection operates through the medium of vehicles. How the replicators make themselves felt in the vehicles, of course, is a matter of empirical enquiry. No one would deny that the effects of the replicators might be quite different in different situations, and combined with other replicators in different vehicles. But, ultimately, what counts is whether or not replicators get passed on. It is in this sense, argues Dawkins -- I think fairly -- that one can speak of replicator selection or, more specifically, of genic selection. The immediate action of natural selection is on the individual organism. The ultimate effect of natural selection is on the individual gene.

Of course, this way of looking at things is still very much what one would expect from a methodological reductionistic strategy. But, I would argue, it is not an objectionably reductionistic way of looking at things. At least, one can say this much: Sociobiological practice is no more offensively reductionistic (in the sense we are now discussing) than is the rest of contemporary Darwinian thought and practice. You might, of course, object to the whole strategy of thinking of evolutionary change in the context of genes, whether these be Mendelian genes or strips of D.N.A. That is another matter. But, once you accept this way of looking at change, then there seems to be nothing especially offensive in sociobiological practice. If you accept the degree of reductionism that the average Darwinian is prepared to countenance and support, then there is no further degree of reductionism to be found in sociobiology, not even when they speak of replicator or genic selection.

But, what if you object to the reductionism of Darwinian theory generally, arguing that the genes are not the right units to think of when speaking of evolutionary change? Here, I am afraid, you will have to introduce your own rival theory first, before we can speak for or against it. What we can say is this. Darwinian theory throws much light on the nature of organic change. Furthermore, the drive to explain things in terms of smaller rather than larger units -- a drive which has been at the heart of the evolution of Darwinism itself over the past century and a quarter -- has brought great insight (Ruse, 1982). Unlike Darwin himself, we now know much about the nature of the units of heredity, and can, therefore, control, predict, and unify with far greater accuracy power than Darwin was ever able to do. And control, prediction and explanatory unification are the end virtues of any science. Thus, again, the defence of methodological individualism considered against Darwinism as a whole is not to be found in commitment to some ethereal metaphysics. Rather, it lies in those very pragmatic factors which always lead one to favour one scientific idea or theory over another.

Hence, I argue without going into the kind of detail which would lead us too far astray, that the development of Darwinism in the past century and more shows that a methodological reductionistic approach has been highly profitable. Therefore, for all their sometimes sloppy language, there is no reason to pick out sociobiologists for special comment and censure.

EPISTEMOLOGICAL REDUCTIONISM

We come now to the third and final stage of our discussion. An epistemological reduction occurs when one part of scientific theory is shown to be a deductive consequence of another part. The two parts, reducing and reduced, could theoretically be part of the same overall subject of inquiry ("intra-field"), or they could theoretically be parts of two rather different subjects of inquiry ("inter-field"). This kind of reduction is, as Ayala rightly notes, something which has been much discussed by philosophers. Supposedly, some of the greatest triumphs of science have involved epistemological reduction, particularly inter-field epistemological reductions (Nagel, 1961; Yoshida, 1977). One thinks, for instance of the derivation of Galilean terrestrial mechanics from Newton's overall physical theory. More recently, one thinks of the derivation of the biological theory of Mendelian genetics from the physico-chemical theory of molecular genetics (Ruse, 1976).

As you might perhaps expect, there has been much discussion in the philosophical literature about how frequently a full-blooded reduction of this kind -- one which involves rigourous deduction -- does actually occur. Some have argued that, usually, what is deduced is something akin to the earlier reduced theory, but not quite the same (Kuhn, 1962). In which case, perhaps one should more strictly speak of a process of "replacement", rather than "reduction". (In both cases, obviously, one might well have a methodological reduction.) It is not necessary for us to enter the details of the debate here. For our purposes, it is enough to take note merely of the general idea behind epistemological reduction, namely that of showing that one science is merely a special instance of, or ready consequence of, another science.

Normally one presumes that, if an epistemological reduction occurs, this involves going from a reducing science, making reference to smaller, more simple entities, to a reduced science, making reference to the more complex. In other words, one presumes that an epistemological reduction normally assumes a methodological reduction. Yet, I am not sure that this is absolutely necessary in theory. One could, perhaps, have a deduction of a theory making reference to entities of one kind, from another theory making reference to entities of much the same order. However, again, these are theoretical questions which need not specifically concern us here. Turning now directly to sociobiology considered in the context of epistemological reduction, questions seem to arise in two different areas (or rather, at two different potential interfaces). On the one hand, we have the prospect of sociobiology as reduced science, in which case the question is that of the potential reducer, and whether there are genuine hopes of strong links between reducer and sociobiology -- links which would be tough enough to warrant the term "reduction" in this context. On the other hand, there is the prospect of sociobiology as reducing science, in which case the question becomes that of the potential reduced science, and whether there are genuine hopes of sufficiently tight links between sociobiology and reduced science to warrant the description "epistemological reduction". Let us consider these two possibilities in turn.

SOCIOBIOLOGY AS REDUCED SCIENCE

If sociobiology (I am including now human sociobiology) is to be reduced to another area of science, we must inquire first into possible candidates within which sociobiology might thus be located as a deductive consequence. Let us start with (what is surely) the most hopeful, namely with candidates promising an intra—field reduction. In other words, initially, let us confine inquiry within the bounds of biological science itself. Now, if one does this, then, since in sociobiology one is dealing with a specific aspect of organic evolution, the most obvious candidate as reducing science is that part of biology which deals most generally with evolution.

But, as we have seen, evolution today is considered to be essentially a function of changes in gene ratios, down through the generations (Dobzhansky et al, 1977).

The area of biology which deals directly with this issue is population genetics. Therefore, the most obvious candidate, as reducing science for sociobiology, is population genetics. If, in fact, an epistemological reduction has either taken place or is possible in principle, it seems most plausible to suggest that it will be centred on the deductive derivation of sociobiology from population genetics.

However, although there are indeed strong links between population genetics and sociobiology, supposing that such links point in the direction of an epistemological reduction fundamentally misrepresents the true relationship between the two sub-areas of biological science. To see this, consider a classic result in population genetics, namely that under certain special circumstances natural selection promotes a balance of different ratios stably within a population. This can come about for a number of reasons, but that worked out in most detail occurs when the heterozygote (that genotype with identical alleles) has a selective superiority over either homozygote (the genotypes with different alleles). (For details, see Ayala and Valentine, 1979; Ruse, 1982).

Now, there is certainly no question of taking this result, and directly deducing something about behavior in populations -- that is to say, directly deducing a sociobiological result. Rather, what one might hope to do is take the theorem about stable heterozygote balance and, in conjunction with a great deal of other information about ecology and behavior, and so forth, infer something interesting and informative about the range of behaviors which a social species might exhibit. One might, for instance, try to show that the combination of aggression and passivity that one encounters is a direct function of the kind of balanced situation that population genetics presupposes. This is valuable and interesting, and there seems to be far more at stake here than a simple replacement. Yet, at the same time, there is hardly enough to warrant a full-blown reduction. More is needed than populational-genetical premises to get sociobiological conclusions.

What then is the correct analysis of the relationship between sociobiology and population genetics, or indeed, between sociobiology and the rest of biological theory? The example just discussed suggests that, in some way, population genetics provides a kind of **background** to the work that the sociobiologist wants to do. In some sense, population genetics is the underlying theoretical substratum on which the work of understanding the biology of social behavior can take place. This analysis certainly fits in with what one finds elsewhere in biology, for population genetics seems to play exactly the same role in many other disciplines! In fact, we find that evolutionary theory has a structure rather like a fan, with population genetics acting as the connective foundation on which all various sub-disciplines of evolutionary theory, like paleontology, biogeography, systematics, and so forth, can depend. Population genetics thus unifies and informs the rest of evolutionary studies (Ruse, 1973).

We see, therefore, that sociobiology now takes its place, along with biogeography, paleontology, and all the rest, as one of the sub-disciplines within the evolutionary family. This, in fact, was precisely the intention of Darwin in **The Origin**, and now in the last twenty years such intentions are being fully realized. Thus, what we have is not a full-blown epistemological reduction, but rather sociobiology being fitted into the background of the overall evolutionary paradigm (Ruse, 1979a). Studying the rest of evolutionary biology, this is precisely what one would have expected. Therefore, in this important sense, there is certainly nothing untoward or threatening about what is going on at this point in the sociobiological realm. One has, if you like, a kind of partial epistemological reduction, but really there is something much fuller than this going on. There is no question of simply elimination everything that is distinctive about sociobiology, and finding it located already in other parts of biology. Instead, sociobiology itself brings its own methods and interests and models and problems to bear on its subject matter. But, this is done and infused against the background of our general evolutionary understanding, particularly as it is represented in modern population genetics.

INTER-FIELD REDUCTION?

The conclusion just established clearly makes moot any question of a full-blown epistemological reduction of sociobiology (even non-human sociobiology) from today's physics and chemistry. Since there is not (and will never be) an intra-field reduction, in the sense normally understood, there will be no inter-field reduction. But, given what has been said already about ontological and mehodological reductionism, might we not someday look towards an epistemological inter-field reduction of the overall Darwinian evolutionary theory, of which sociobiology is a part, to the theories of physics and chemistry? Indeed, given what has been acknowledged already, about the reduction of Mendelian genetics to molecular genetics, is not such a reduction already on its way?

Since we are rather peering into the future at this point, one can hardly hope to give definitive replies to these (admittedly interesting and important) questions. My own sense is one of caution, although one of optimistic caution (if that is the appropriate predicate). On the one hand already we see the effects and importance of physico-chemical understanding coming through into evolutionary biology, even through to sociobiology (Lumsden and Wilson, 1981). The work that molecular biologists have done on the structure of the units of heredity tell us much, for instance, about variations in populations, and this is now something which is being picked up by sociobiologists. For instance, work is now going on apace even in the human species, showing how different races might have different behavioral dispositions because of different molecular sub-strata. One would expect that molecular understanding of the organism, even as it reflects through to behavior will become more and more important, both as molecular biology is itself developed, and as sociobiologists extend the scope of their own understanding.

On the other hand, it is one thing to say that molecular understanding is and will always be very important for evolutionists. It is another to say that, at some point, Darwinian evolutionary theory (including sociobiology) will consist exclusively of molecular-type explanations, which would seem to be the ultimate limit of an epistemological reductionistic stance. It is even more to say -- what a full inter-field reduction requires -- that Darwinian theory will or could be explained from physics and chemistry as presently construed. Apart from pragmatic factors -- think, for instance, how complex would even be a molecular explanation of hymenopteran behavior, let alone humans -- I suspect there are theoretical reasons why a full-blown epistemological reductionistic programme absorbing Darwinian theory, could never be realized. And this applies both to such a programme using physics and chemistry in their present state, and to such a programme relying on a physics and chemistry much extended but of the same type as exists today.

The main reason for doubt of this kind lies, most obviously, in the existence and nature of consciousness. Thought and action may not demand special life forces or substances (i.e. may not demand a break with ontological reductionism), but this is not to say that they require no understanding not furnished by today's talk about molecules. A thought is not merely an electron buzzing around -- at least, a thought requires a different kind of understanding than simply information about mass, velocity, and the like. Therefore, inasmuch as evolutionary theory includes consciousness as one of its areas of study -- and in including human sociobiology it certainly does this -- it seems to demand of a potential reducing science, some say of going beyond blind forces and fields and the like. And this seems to demand going beyond physics and chemistry as they exist at the moment.

But, even if we drop consciousness from the field of evolutionary inquiry, hopes of epistemological reduction seem remote. This reason lies in the peculiarly forward-looking or teleological nature of the organic world (Ruse, 1977). What is truly distinctive about organisms, and what biological understanding -- particularly Darwinian biological understanding -- picks up is that organisms are not simply thrown-together, random collections of entities, rather like patterns in the sand by the shore. Rather, organisms are integrated, functioning beings. They work, that is to say, parts of organisms are directed towards an end, namely, that of survival and reproduction. In short, organisms have "adaptations".

It is for this reason that it is appropriate, when speaking of organisms, to use terms like "function", as in: "What function does the sail on the back of the dimetrodon serve?" Here, one is picking out adaptations. Such function-language is quite inappropriate in the inorganic world. No one, for instance, would ask what function the moon serves. It is not adapted. I believe (and indeed have argued in detail elsewhere) that the functional language of biology points to an ingrained difference in biological understanding, from any which is to be found in the physico-chemical sciences. Biological objects demand a forward-looking or teleological type of understanding which inorganic objects do not (Ruse, 1981). And, as just noted, this functional type of understanding is something which is pervasive, not only in pre-evolutionary biological theories, but also most particularly in Darwinian (and neo-Darwinian) evolutionary theory. Darwin himself was fully aware of the need to explain adaptational functioning. Furthermore, through his theory of natural selection, he felt with good reason that he had succeeded magnificently (Ruse, 1979b). Adaptation is not the immediate result of God's Design, but of the natural process of differential reproduction.

Focussing on our own specific topic of interest, sociobiology, since this subject is part of Darwinian theory, one expects to find functional type explanations endemic in sociobiology. And this, one does indeed find. Sociobiologists, no less than the rest of the biological community, ask questions about ends or design or purposes. Thus, for instance, one can ask: "What function does a certain mating behaviour, like the aggressiveness of the male red deer, serve?" Or: "What function does the sterility of the hymenoptera workers uphold?" And so forth (King's College Sociobiology Group, 1982). If one attempted explanations purely in terms of molecules, then, unless one had a drastic reorganization of the kind of understanding that presently holds in physics and chemistry, one would lose an essential aspect of biological (including sociobiological) understanding. Today's physics and chemistry are not teleological.

In short, I see no ready prospect of an inter-field epistemological reduction of evolutionary theory (including sociobiology), and suspect there are theoretical reasons why today's type physics and chemistry could never serve as adequate reducing sciences. But, I do hasten to add that this is all very tentative. Moreover, I certainly do not preclude the possibility of a revised physico-chemical theory (or, perhaps. an augmented theory) in principle dealing with both mind and teleology (assuming that these are quite distinct). Perhaps appropriate translation principles will take us from the innanimate to consciousness, rather as appropriate principles take us from molecules in motion to temperate and pressure (in gas theory). Perhaps a physics and chemistry of the organism could be teleological. Today's molecular biology suggests this might be possible, since it already incorporates such design-like concepts as the genetic "code" (Ruse, 1977). Hence, I am not saying that a methodological reductionistic strategy could not in theory take us all the way to physico-chemical understanding, even of conscious behavior. But this would certainly be no epistemological reduction starting only with today's physico-chemical theories.

SOCIOBIOLOGY AS REDUCING SCIENCE

We turn now to the other potential interface, namely between sociobiology and other sciences. Sociobiology is now the would-be reducer. Here, most obviously, we look outwards to the social sciences. The interface, if there is to be one, will be between human sociobiology and such subjects as anthropology, sociology, economics, and the like. (I assume no argument is necessary to exclude subjects like geology!)

Now, obviously, the situation here is rather different from the interface between the rest of evolutionary theory and sociobiology. We are not, at the moment, dealing with a situation which is already at least partially articulated. Sociobiology already draws on the rest of biology, just as biology in turn already draws on the physico-chemical sciences, as it does in molecular biology. What we are faced with now is more the prospects of human sociobiology at some point interacting in some way with the work of the social sciences (Wilson, 1978). The question is whether (epistemologically speaking) it is to be one of reduction, replacement, or something in between.

Something has to happen. It will not be possible for sociobiology (speaking now particularly about human sociobiology) and the social sciences to ignore each other indefinitely. Indeed, already they draw together, attempting in their various ways to account for similar, nay identical, phenomena. One thinks, for instance -- to take but one instance -- of the attention paid to the widespread existence of those incest barriers which exist in the human species. From culture to culture, virtually universally, one finds that there are barriers, often backed by explicit taboos between intra-familial sexual relationships. Take sibling incest. Even in societies where sexual practice is fairly loose and casual, brothers and sisters simply do not copulate. Furthermore, even in those few societies where incestuous sibling relationships have been allowed or, indeed, made obligatory, such relationships tend to be extremely limited and due to special rather artificial circumstances. Certainly, the number of exceptions is far, far, outweighed by the universality of incest barriers (van den Berghe, 1983).

Both sociobiologists and more conventional social scientists have been intensely interested in sibling incest. Both try in their ways to explain and account for it, so interaction is inevitable. And, indeed, this kind of incest is a place where territorial disputes are already starting to occur. Hence, it seems an ideal point on which we can hang our discussion. What sort of future relationship does an example such as this presage for human sociobiology and the social sciences? Let us look in a little more detail at the actual explanations that sociobiologists and social scientists offer for sibling incest barriers. Then, perhaps, we will be in a better position to answer such a question as this.

The sociobiological explanation of sibling incest barriers is straightforward. Although there is generally a high premium on reproduction -- this being a natural consequence of any evolutionary perspective which makes success in the struggle for reproduction the prime causal factor -- sociobiologists argue that (sibling) incestuous relationships are counter-productive, because they lead to horrendous inbreeding effects (van den Berghe, 1983; Wilson, 1978; Alexander, 1979). Simply put, the offspring of very close relatives tend to be biologically unfit, and thus not well-suited for life's struggles. Hence, there is a strong selective pressure against sibling relationships. Thus, there has evolved a disinclination to reproduce with close relatives. In terms of the Wilson-Lumsden theory of epigenetic rules, it is argued that we have an innate disposition not to have sex with siblings. "The evidence suggests the existence of a genetically-based bias curve in which the preference for outbreeding as opposed to incest is very strong" (Lumsden and Wilson, 1981, p.86).

Turning next to the social sciences, the situation is somewhat more complex, for a number of different explanatory models have been put forward to account for incest barriers. One of the earliest, dating back to the last century, was precisely that of the sociobiologists! The anthropologist Edward Westermark (1891), argued that incest barriers are put in place for the prevention of deleterious biological effects. What the sociobiologists have done is simply picked up on Westermark's explanation and set it within the context of modern biology, particularly within our modern knowledge of genetics and its effects. For the moment, no more need be said about this position.

Coming down into this century, at least three different approaches have been taken toward (sibling) incest barriers. First, there is the Freudian psychoanalytic account. It was suggested by Freud (1913) that incest barriers date back to human pre-history, when a group of brothers ganged together to kill their father. In order to prevent strife between themselves, they decreed that sexual relations within the family would be taboo. What Freud believed was that these actual actions were, in some sense, imprinted upon the human collective consciousness, and thus have been passed down through the generations. What was once a conscious decision is now part of our biology.

Second, we have the well-known explanation of the French anthropologist, Claude Lévi-Strauss (1969). He argues that incest barriers are a function of the devices used by males (at least, in pre-literal societies) to promote harmony between different tribes. According to Lévi-Strauss, cooperation, relationships, treaties, and so forth, between otherwise warring tribes, can be avoided by the exchange of females. Thus, argues Lévi-Strauss, there is a good reason not to use up one's own women for oneself, but rather to keep them pure and intact for the bartering process. As in so much else, we in industrial societies carry the social legacy of our ancestors.

Third, we have the recent work of a number of cultural anthropologists who have made fairly detailed studies of situations in which incest barriers might or might not be put into place. Particularly, these social scientists have looked at situations where the distinction between social sibling and biological sibling gets confused. For instance, on the Israeli kibbutz, biologically unrelated children are brought up as members of the same family. What these anthropological students of human behavior have found is that, although there is no biological relationship, children thus raised together -- "having shared the same potty" -- erect between themselves, subconsciously, incest barriers. They behave as though they were biological siblings. Hence, conclude these social scientists, the reason for incest barriers lies in a kind of negative imprinting (Shepher, 1971; 1979). We are unable to relate sexually to those with whom we were raised, as if in the same family. Normally, of course, social siblings and biological siblings are one and the same. Hence, we get the fact that it is (biological) brothers and sisters who do not want to breed together.

RIVAL EXPLANATIONS

Now, how does the sociobiological explanation of sibling incest barriers relate to the four social science explanations? As noted, the sociobiological explanation and Westermark's explanation are one and the same. In fact, obviously, the sociobiologists were much influenced by Westermark in their own work. Clearly, therefore, at this point we get a fairly straightforward fusing of sociobiology and social sciences, although it is hardly a full-blooded case of epistemological reduction. There is less a question of deriving one science from another, and more a question of moving the barriers on what you are to count within the one science and what you are to count within the other. What was hitherto part of social science, now, according to the sociobiologists, must at least equally be counted as part of the biological sciences. I take it that a move of this kind causes no conceptual confusions or particularly interesting problems, although obviously sociologically there may be tensions. At least part of the objection to sociobiology in recent years has come precisely because social scientists feel they are under threat of being swallowed up by the biological sciences (eg. Sahlins, 1976). Whether the swallowing takes the form of deductive derivation, or simply of moving barriers, is presumably a matter of little moment to those who feel that they are losing out.

Moving next to the Freudian explanation, clearly the sociobiological explanation cannot coexist harmoniously together -- one or other must be wrong. If incest barriers are due to patricide, then they are certainly not due to the deleterious effects of close inbreeding. Today, anyone who has any awareness of the biological sciences, whether or not he opts for the sociobiological alternative, agrees that Freud's answer is seriously flawed. The reason for this is that Freud relies on an illegitimate notion of inheritance (Sulloway, 1979). He was, in fact, much influenced by the ideas of Lamarck, believing that it is possible for some feature to be ingrained on the adult form or phenotype, and that this can then be transmitted into future generations directly through the sex cells. All the evidence that we have, not just sociobiological, points decisively against this Lamarckian idea of the inheritance of acquired characteristics (Ruse, 1982). There is simply no feed-back mechanism, from the developed organism to its sex cells, and hence to its offspring. Therefore, if we do in fact assume the truth of the sociobiological account, then what we face is not epistemological reductionism, but epistemological replacement. The old science is being removed, and the new is moving in. But do note that, even if we do not accept sociobiology, there is no reason to accept Freud. Hence, whatever does happen will not be because of some ethereal commitment to reductionism or whatever, but simply because an enterprising hypothesis is false. Freud is not inadequate metaphysics -- it is disproved science. (The reader might want to look at Fox [1980], which is an enterprising attempt to pull Freud up to the standards of modern biology, and then relating his insights to the incest question. In Ruse [1979a], I look at various explanations for homosexual orientation, arguing that Freud and sociobiology might mesh harmoniously.)

Turning third to the Lévi-Strauss explanation, where incest barriers are taken to stem from the virtues of exchanging women between tribes dominated by men, the sociobiological/social science relationship seems **prima facie** that just encountered in the Freudian case. At least one of the putative explanations must be false. If Lévi-Strauss is right, then the main reason for incest barriers has little to do with the ill effects of close inbreeding. Conversely, if the·sociobiologists are right, then Lévi-Strauss is either wrong or irrelevant. Hence, either way, we do not seem to have here the prospect of reduction. If Lévi-Strauss is wrong, then presumably sociobiology is simply going to replace his type of thinking. On the other hand, if sociobiology is wrong, then at least some revision is going to be required in the biological theory.

However, there is a difference between this case and the Freudian case, in that here the social science explanation is not known to be quite definitely false. Perhaps it is true! In which case, an interesting question arises. Supposing Lévi-Strauss is right and sociobiology is wrong, does this mean that sociobiology as such is irrelevant; or, rather, that the current sociobiological models are inadequate? There are at least two possible options here, both of which are compatible with some form of Lévi-Strauss's hypotheses. On the one hand, one might argue that everything happens between tribes, the actions of the men, and of the women is purely cultural, with no effects at all. on the genes, and with no connections to biological fitness. Men exchange women, and this has nothing at all to do with their biology. There are no genetic causes o effects, whatsoever. In which case, at the very least, one would want to say that sociobiology is totally irrelevant. An attempt at reduction/replacement (it does not matter which) was tried, and failed.

On the other hand, there is surely the possibility of entertaining some kind of refined or revised sociobiological theory: one which does take Lévi-Strauss's hypotheses extremely seriously. In short, you have your sociobiological cake, and eat it too! One might argue that the important thing is that humans reproduce as efficiently as possible. Those males who enter into relationships with others, exchanging their females, get a bigger biological pay-off than those who simply use the closest available females, namely their sisters. In other words, it is a biologically preferable form of behaviour to give up your sister and take somebody else's sister, than simply to copulating with your own sister. The reason for this being that the immediate benefits of sibling reproduction are far outweighed by the dangers of attack from other tribes. Thus, a small amount of effort now, in putting off reproduction. pays magnificent bonuses later on.

Note that this argument need not in any way be a group selection explanation, because, because one can argue that the individual male thus gains. And, lest this all sounds too sexist to be true, one could argue also that it is in the females' self-interest to be passed on to other males, rather than breeding simply with closest relatives. With inter-tribe sister exchange, females and their offspring are, no less than men, likewise thus freed from the terrors and dangers of attack from without.

I hasten to add that I am not saying that any of this just-offered line of argument is true. What I am saying is that I see no reason why a sociobiologist should not adopt such an explanation as this, and claim it as part of his own. (See van den Berghe [1979] especially, for sympathetic responses to his teacher, Lévi-Strauss.) But here, obviously, we would have neither replacement nor (epistemological) reduction by sociobiology. Rather, established social science is helping the sociobiologist to devise his theories, which latter can then put social science in a new and deeper light. Why would women put up with exchange? (And do not say: "Because they have no choice." It is hardly true, and ignores the underlying problem of why women rather than men are of a kind to be exchanged.)

Even when the sociobiology is finished, we have neither replacement nor reduction. The social science remains, so there is no replacement. And the social science is hardly a deductive consequence of biology. So there is no (formal) reduction. We have, instead, an amalgam of biological and social science, with the former giving the background and the latter setting the problems and giving the details. In other words, we have the kind of relationship that obtains between population genetics and sociobiology itself. We certainly do not have the kind of threatening encounter feared by so many social scientists.

Which brings us to the third and final explanation of this century, by social scientists, for incest barriers. This is the explanation which makes reference to the supposed negative imprinting which occurs between close relatives, when they are raised together. Here, unlike (say) the Freudian explanation, we seem less to be looking at the ultimate causes of why incest barriers fall into place (in the Aristotlean sense, the final causes), and more at the immediate causes (or, in the Aristotlean sense, the proximate causes). The question now is not why incest barriers come into place; but, rather, what makes them come into place. And, the answer given is negative imprinting.

But, such a negative imprinting explanation could easily be slotted into a number of wider ultimate theories, including most particularly the currently-held sociobiological theory. One denies that there are genes for sibling incest, per se. Indeed it is difficult too know quite how such genes as these could come into effect, except perhaps through pheromones or some such chemical kind of detection. Rather, one asserts there is a biological premium on not copulating with close relatives, and that the way in which this premium is cashed is through a general distaste for such close inbreeding with anyone with whom one grew up in childhood. The genes, therefore, are not against sleeping with brothers and sisters, but against sleeping with anyone with whom one was raised as a social sibling. Of course, normally social siblings and biological siblings are one and the same, so this means that the genes achieve their ends in the long run, anyway. But it is possible, either accidentally or through human manipulation, to fool the genes. In such cases as the Israeli kibbutz, this is precisely what happens.

The logical situation here is certainly not one of replacement between sociobiology and social sciences, nor yet does it seem to be one of straight epistemological reduction. One is not deducing the negative imprinting from a theory about the genes, and so forth. Rather, we seem to have much the situation which I supposed would occur if one accepted Lévi-Strauss's work and tried to give it a sociobiological backing. Sociobiology stets the outer parameters, as it were, and then the social science explanation fits in the particular details of how the actual mechanisms are to go into effect. Sociobiology tells us that there will be a genetic predisposition towards incest avoidance, but it says nothing in itself about how this predisposition is to got into place. The work of the social scientists, however, show us precisely how this works. We thus have a fusion from two hitherto-separate fields of inquiry, rather than the triumph of one over the other. No one is pushed aside or made redundant. (See Ruse, 1979a, for a similar analysis of various explanations for homosexual orientation.)

All in all, therefore, considering the sociobiology/social science interface, there is no easy and straight answer to the relationship as it exists at the moment, or as it might exist in the future. That sociobiology and the social sciences are already coming into contact, and will continue to do so more in the future, seems indisputable. But, will this be one of replacement, reduction or something else? In part, certainly it seems that replacement must occur. Some of the sociobiological explanations cannot be held simultaneously with some of the social science explanations. However, it does not follow that when sociobiology and the social sciences come into conflict it must necessarily always be the social sciences that give way, and sociobiology which triumphs. The truth, I suspect, is very much an empirical question. If, for instance, anthropological practices are found to fit far more with what Lévi-Strauss argues than with what the neo-Westermarkians argue, then whatever else, sociobiology must give way in significant respects. A replacement of the (currently held) biology will occur.

But, what happens when the encounter between biological and social sciences is more harmonious? Should we then look for a reduction, of the epistemological variety? Is this what a good scientific attitude demands? The example of incest barriers suggests that this is not the case at all. There is anything but -- and anything but the need for -- a simple epistemological reduction, with the social sciences being deduced straight from sociobiological principles, even human sociobiological principles. There will be a merging, which will probably be as much a matter of redefining barriers and boundaries, as of actual derivations. Even more, one looks to the kind of fusing that we saw occurs between sociobiology and the rest of biology. What we should expect to find is that sociobiology will provide the overall background, the ultimate causes, against which specific social science explanations can be fitted. The social sciences will give us the mechanisms and information, and then a knowledge of our Darwinian heritage will fit this information into an allembracing picture. In such a case as this, there will be no losers, and only winners.

CONCLUSIONS

There are important questions of reductionism which surround the sociobiological enterprise. However, it must now be agreed that simply speaking of sociobiology, including here human sociobiology, as "unduly reductionistic", confuses the issue more than it clarifies. One must first distinguish between the senses of reductionism, and decide on the particular sense that one is using in a certain context. When one has done this, one can see that in certain of these various senses of reductionism, although sociobiology is undoubtedly reductionistic, it is no more so -- and certainly no more guiltily so -- than virtually any other branch of science today.

Ontologically speaking, no one would want to fault sociobiologists for excluding vital forces, and few today think that invoking things like Cartesian consciousness is a particularly worthwhile scientific strategy. Methodologically speaking, it is true that sociobiologists have sometimes been guilty of confusing issues, through rather strident reductionistic claims. However, here again the actual commitment to reductionism by sociobiologists is not in itself necessarily wrong. Sociobiologists do not make some illegitimate metaphysical commitment to methodological reductionism, and then ride roughish over the facts of experience, in order to bring their programme to successful and triumphant conclusion. Rather, the sociobiologists' commitment to methodological reductionism is part of a well-articulated and fruitful scientific attitude. When dealing with social behavior, there are good reasons for being reductionists in this respect. Furthermore, these reasons have paid ample dividends in the past. Hence, it is only good scientific practice to assume that the strategy is one worth continuing. This is not to say that methodological reductionism must always be the right approach to dealing with social behavior from an evolutionary perspective: but, as we have seen, even sociobiologists admit this.

Finally, speaking of epistemological questions to do with reductionism (in the sociobiological context), we have learned that there is little prospect or apparent virtue in hard-line reductionism. It is undoubtedly true that sociobiology itself looks towards a fusion with the rest of biological theory. Indeed, unless sociobiology is itself firmly connected with the rest of evolutionary theory, it makes little sense. In like manner, turning towards the social sciences, we see at least prospects of a mutually profitable relationship occurring here. This is a relationship which will fuse biological and social science. But, at neither interface does one anticipate the deductive derivations required for epistemological reduction. There will be far more give and take. To speak of sociobiology as "reductionistic" -- especially, to speak of sociobiology as "unduly reductionistic" -- is to miss the point entirely.

I emphasized at the beginning of my discussion, that, at most, this inquiry into sociobiology and reductionism can only be a prolegomenon to a full evaluation of the worth of sociobiology, including human sociobiology. Deliberately, I have avoided questions to do with truth, falsity, ill-foundedness, and so forth. Without some answers to these, no full evaluation is possible. My aim has been, deliberately, merely one of conceptual clarification. But, given criticisms which have been levelled, this limited inquiry has paid dividends. In important respects, sociobiology is indeed reductionistic. However, there is no warrant whatsoever for arguing that this attitude is, in itself, a fault. To the contrary, if anything, sociobiology shows why reductionism, properly understood and used, has been such a powerful force in the forward progression of the scientific enterprise.

Bibliography

Alexander, R. 1979. **Darwinism and Human Affairs**. University of Washington Press, Seattle, Wash.

Ayala, F.J. 1974. Introduction. Pages vii–xiv in F.J. Ayala and Th. Dobzhsansky eds. **Studies in the Philosophy of Biology**. Macmillan, London.

Ayala, F.J. and J.W. Valentine. 1979. **Evolving**. Benjamin/Cummings, Menlo Park, California.

Barash, D.P. 1982. **Sociobiology and Behavior**. 2nd ed. Elsevier, New York.

Brandon, R. and R. Burian. 1984. **Genes, Organisms, Populations**. MIT Press, Cambridge, Mass.

Churchland, P.M. 1984. **Matter and Consciousness**. MIT press, Cambridge, Mass.

Darwin, C. 1859. **On the Origin of Species**. John Murray, London.

————————. 1871. **Descent of Man**. John Murray, London.

Dawkins, R. 1976. **The selfish Gene**. Oxford University Press, Oxford.

————————. 1978. Replicator selection and the extended phenotype. **Zeitschrift für Tierpsychologie**, 47, 61–76.

Dobzhansky, Th., F.J. Ayala, G.L. Stebbins, and J.W. Valentine. 1977. **Evolution**. Freeman, San Francisco, Calif.

Doolittle, W. and C. Sapienza. 1980. Selfish genes, the phenotype paradigm, and genome evolution. **Nature**, 284, 601–3.

Fox, R. 1980. **The Red Lamp of Incest**. Dutton, New York.

Freud, S. 1913. **Totem and Taboo**. In **Collected Works of Freud** (ed) J. Strachey, Vol. 13, 1953. Hogarth, London.

Gould, S.J. 1980. Caring groups and selfish genes. In **The Panda's Thumb**. W. W. Norton, New York, 85–92. Reprinted in E. Sober ed **Conceptual Issues in Evolutionary Biology**, pages 119–124. MIT Press, Cambridge, Mass.

Hempel, C. 1966. **Philosophy of Natural Science**. Prentice-Hall, Englewood Cliffs.

Hamilton, W.D. 1964. The genetical evolution of social behaviour. I. **Journal of Theoretical Biology**, 7, 1-16.

————————. 1964. The genetical evolution of social behaviour. II. **Journal of Theoretical Biology**, 7, 17-32.

King's College Sociobiology Group eds. 1982. **Current Problems in Sociobiology**. Cambridge University Press, Cambridge.

Kuhn, T.S. 1962. **The Structure of Scientific Revolutions**. University of Chicago Press, Chicago, Illinois.

Lévi-Strauss, C. 1969. **The Elementary Structures of Kinship**. Beacon Press, Boston, Mass.

Lewontin, R.C. 1974. **The Genetic Basis of Evolutionary Change**. Columbia University Press, New York.

————————. R.C. 1977. Sociobiology: a caricature of Darwinism. In F. Asquith and F. Suppe eds **PSA, 1976**. Philosophy of Science Association, East Lansing, 2, 22-31.

Lewontin, R.C., S. Rose, and L.J. Kamin. 1984. **Not in Our Genes: Biology, Ideology, and Human Nature**. Pantheon, New York.

Lorenz, K. 1952. **King Solomon's Ring: New Light on Animal Ways**. Methuen, London.

————————. 1966. **On Aggression**. Methuen, London.

Lumsden, C.J. and E.O.Wilson. 1981. **Genes, Mind and Culture: The Coevolutionary Process**. Harvard University Press, Cambridge, Mass.

————————. 1983. **Promethean Fire**. Harvard University Press, Cambridge, Mass.

Maynard Smith, J. 1972. Game theory and the evolution of fighting.In **On Evolution**, Edinburgh University Press, Edinburgh.

————————. 1974. The theory of games and the evolution of animal conflict. J. **Theoretical Biology**, 47, 209-21.

Mayr, E. 1975. The Unity of the Genotype. **Biologisches Zentralblatt**, 94, 377-88.

————————. 1982. **The Growth of Biological Thought**. Harvard University Press, Cambridge, Mass.

Nagel, E. 1961. **The Structure Science**. Routledge and Kegan Paul, London.

Popper, K.R. and J.C. Eccles, 1977. **The Self and Its Brain**. Springer–Verlag, New–York.

Rosenberg, A. 1985. **The Structure of Biological Science**. Cambridge University Press, New York.

Ruse, M. 1973. **The Philosophy of Biology**. Hutchinson, London.

–––––––. 1976. Reduction in genetics. Pages 653–70 in R.S. Cohen et al eds, PSA **1974**. Reidel, Dordrecht.

–––––––. 1977. Is biology different from Physics? In R. Colodny ed. **Laws, Logic, Life. Pittsburgh Studies in The Philosophy of Science**. Pittsburgh University Press, Pittsburgh.

–––––––. 1979. **Sociobiology: Sense or Nonsens**. Reidel, Dordrecht.

–––––––. 1979. **The Darwinian Revolution: Science Red in Tooth and Claw**. University of Chicago Press, Chicago.

–––––––. 1980. Charles Darwin and group selection. **Annals of Science**, 37, 615–30.

–––––––. 1981. The last word on teleology, or optimality models vindicated. Pages 85–101 in **Is Science Sexist? And Other Essays on the Biomedical Sciences**. Reidel, Dordrecht.

–––––––. 1982. **Darwinism Defended: A Guide to the Evolution Controversies**. Addison–Wesley, Reading, Mass.

Sahlins, M. 1976. **The Use and Abuse of Biology**. University of Michigan, Ann Arbor.

Shepher, J. 1971. Mate selection among second generation kibbutz adolescents and adults. **Arch. Sexual Behavior**, 1(4), 293–307.

––––––––. 1979. **Incest: The Biosocial View**. Harvard University Press, Cambridge, Mass.

Sober, E. 1984. **Conceptual Issues in Evolutionary Biology**. MIT Press, Cambridge, Mass.

Sulloway, F.J. 1979. **Freud: Biologist of the Mind**. Basic Books, New York.

Tinbergen, N. 1953. **Social Behaviour in Animals**. Methuen, London.

––––––––. 1953. **The Herring Gull's World: A Study of the Social Behaviour of Birds.** Collins, London.

Trivers, R.L. 1971. The evolution of reciprocal altruism. **Quarterly Review of Biology, 46,** 35–37.

van den Berghe, P. 1979. **Human Family Systems.** Elsevier,' New York.

van den Berghe, P. 1983. Human inbreeding avoidance: culture nature. **The Behavioral and Brain Sciences, 6,** 91–124.

Westermark, E. 1891. **The History of Human Marriage.** Macmillan, London.

Williams, G.C. 1966. **Adaptation and Natural Selection.** Princeton University Press, Princeton.

Wilson, E.O. 1975. **Sociobiology: The New Synthesis.** Harvard University Press, Cambridge, Mass.

––––––––. 1978. On **Human Nature.** Harvard University Press, Cambridge, Mass.

Yoshida, R. 1977. **Reduction in the Physical Sciences.** Dalhousie University Press, Halifax, N.S.

Percy Löwenhard

THE MIND-BODY PROBLEM: SOME NEUROBIOLOGICAL REFLECTIONS

1. THE ONTOLOGICAL BACKGROUND FACTORS

The human brain is one of the most marvellous achievements of nature. It is not only the most complex structure in that part of nature which is known to us, but its complexity includes the existence of functional principles whose details to a large extent are still more or less enigmatic.

George Wald once typified the essence of it:

"Life has a status in the physical universe. It is part of the order of nature. It has a high place in that order, since it probably represents the most complex state of organization that matter has achieved in our universe. We, on this planet, have an especially proud place as man; for in us as men, matter has begun to contemplate itself." (George Wald, 1960 – 61; c. fr. Gunn, 1972, p. 131).

The subject matter of brain and consciousness reflects a multi-levelled complexity and conceptual elusiveness. While there is knowledge based on well established facts, there are also phenomena which are difficult to understand and it is even not always possible to state these difficulties in a precise and unambiguous language. This is also reflected in the variety of ways in which the topic is treated in textbooks and encyclopaedias. More extensive examples are given by Löwenhard (1981).

It is evident that complexity is the key issue in this connection. As David Hubel (1979) bluntly puts it: "The incredible complexity of the brain is a cliché, but it is a fact". The brain not only utilizes in the order of 10^{11} neurons and 10^{15} synoptic connections, but the pattern of connections is subject to alterations during information processing and learning, which provides the necessary basis of the brain's functional plasticity, some details of which will be given later on.

P. Hoyningen-Huene and F. M. Wuketits (eds.), Reductionism and Systems Theory in the Life Sciences, 85–135.
© *1989 by Kluwer Academic Publishers.*

Any description of such systems has to take the above mentioned facts into consideration. There seem to be at least three concepts which are central in this task. One is the notion of hierarchically organized systems with dynamic interaction; the second one is the concept of evolution, both in the Darwinian sense as phylogenesis and with respect to the individual as ontogenesis. The third one, finally, is the view, that the adaptation of a system represents an information process. These three aspects of living systems are integrated within the rapidly developing thought model of evolutionary epistemology (see e.g. Campbell, 1974; Riedl, 1979; Vollmer, 1981; Lorenz & Wuketits, 1983). Before entering into details, some ontological and epistemological statements should be made explicit as a background to what is said later on. This is necessary, since at least in psychology they do not conform to all theory.

1.   There exists an external reality, independent of an observer.

2.   This external reality, our universe, forms a whole, which at least partially is ordered.

3.   Knowledge about this order is at least partially available.

4.   Living organisms are part of the external world and may act as observers.

5.   Information about the external world is stored within living organisms. At the human level, the following may be added: independent of the complexity of the technique of observation and the degree of sophistication of an external processing of information, there is no knowledge until the information has been interpreted and made conscious by the brain of an observer.

Phenomena such as "life", "consciousness", "matter", "energy", "space" and "time" must be regarded as basic attributes of our universe. The distribution of matter and energy as a function of time may then be described in terms of entropy, where cosmological entropy is expressed as a ratio between the amount of radiation (photons) and particles with mass.

Physical theory ultimately describes the nature of our world in terms of particles and interaction between them. The term "particle", however, must not be interpreted in the sense of everyday language but as a quantum mechanical concept with both "wave" and "particle" features. While there are four known types of interaction (or forces), only the electromagnetic interaction is of interest here (with gravitation at the second place), since it determines the majority of phenomena within the mesocosmic range of dimensions: the coherence of matter, the molecular bond, chemical reactions, electric and magnetic phenomena and all their applications such as the transmission of energy and information.

"Life" and "consciousness" nevertheless represent a higher ontological level of phenomena (within the same basic ontology), where "consciousness" presupposes the existence of "life". The emergence of living systems is dependent on the ability of matter to organize itself into complex structures under suitable conditions. One critical factor is the existence of a universe which has a sufficiently low level of entropy and allows for local variations of it. An example of this is the fact that the evolution of "life" on earth is made possible by the emission of low entropic energy at 300 K. In this sense "life" and "consciousness" may be said to exist as potentialities of our universe, while their emergence as actual phenomena is dependent on the existence of favourable local conditions.

Some comments should finally be made on the concepts of energy and entropy. Energy is a basic attribute of both matter and radiation, but it is not known if the phenomenon exists independent of them. Mainly, energy is recognized by its effects; one of them is the ability to perform work. At the time of Aristotle, the term *energeia* meant activity, force, power, the cause of events or action, i.e. what we today would call its effects. Both in everyday language and psychology, this still is implied by the meaning of the term, but there is no precise and independent physical definition.

The term energy was introduces in 1807 by Thomas Young to denote half the amount of "vis viva" = "living force" = mv^2. In a more general sense, the term was used by W.J.M. Rankine in 1854.

If we speak of the mass–energy relationship, this means at the level of elementary particles that the annihilation of a particle of mass m gives rise to radiation of frequency f (or the photon energy E)

$$E = h.f = m.c^2_0$$

The equation says that a quantity of energy (E) is equivalent to an quantity of matter (mass m). Energy then is an attribute of radiation. The factor c^2_0 is a constant of proportionality, the "exchange rate" between mass and energy.*)

Energy is known to exist in different forms. We speak of kinetic energy of moving matter, of potential energy of a field (its ability to perform work), chemical energy, electrical energy, atomic energy or heat (basically kinetic energy of

---

*) E= energy in Joule (J), including an initial kinetic energy $E_k = 1/2 \ mv^2$

   m = mass in kg

   c_0 = velocity of light in vacuum ($2.997.10^8$ m/s)

   h = Planck's constant ($6.625 . 10^{-34}$ Js)

molecules). Forms of energy can be transformed into each other, but energy can neither be destroyed nor created (if one includes mass as an equivalent of energy). This essentially is the content of the First Law of Thermodynamics.

The definition of energy as "the ability to perform work" is only partially true. It applies, with respect to heat, to that part of energy which Z. Rant at the beginning of the 1950s has named exergy and which formally is defined as the product of negentropy and the absolute temperature (T) of the environment. Negentropy here is defined as the difference between the actual and the maximum entropy of the system (Nordling, 1982).

Now, the notion of entropy has been the issue of some debate. At an ICUS conference, physicist Alvin Weinberg told the following anecdote: When Claude Shannon in connection with his Theory of Communication introduced the quantity H as a measure of uncertainty or information, John von Neumann is supposed to have said: "Call it entropy, for nobody really knows what entropy is". Nevertheless, the term entropy appears in both thermodynamics, statistical mechanics and information theory. The concept was originally introduced by Rudolf Clausius in 1865 as the rate of the heat content and absolute temperature (Q/T) of a physical system in order to characterize the possible direction of a process within energetically closed systems.

In terms of this principle, the Second Law of Thermodynamics says: "In a closed system the entropy of the system either increases or remains constant":

$$dS \geq 0 \quad (dS = \text{change of entropy})$$

where $dS = 0$ denotes reversible processes and $dS \, \grave{} \, 0$ irreversible processes. All natural processes are irreversible.

Any transformation of energy into other forms has the consequence that a part of it gets "lost" in the sense that it is no longer available for useful work. The entropy increases and exergy decreases (for details see e.g. Sears, 1953; Fermi, 1956).

2. LIVING SYSTEMS

"Life" is a common label which we apply to the totality of organisms or living systems which constitute the biosphere of earth. All living systems are characterized by a set of common and individual characteristics. The functional properties which are necessary to sustain the continuous existence of a living system may be called its life functions. Organisms are a subset of living systems and are characterized by a particularly high degree of autonomy. Presently the following four characteristics are viewed as sufficient conditions for life:

1. Metabolism, 2. Reproduction, 3. Mutability 4. Interaction between functional elements (e.g. proteins) and carriers of information (DNA or RNA).

According to Kaplan (1978), these features may be summarized in a simple scheme:

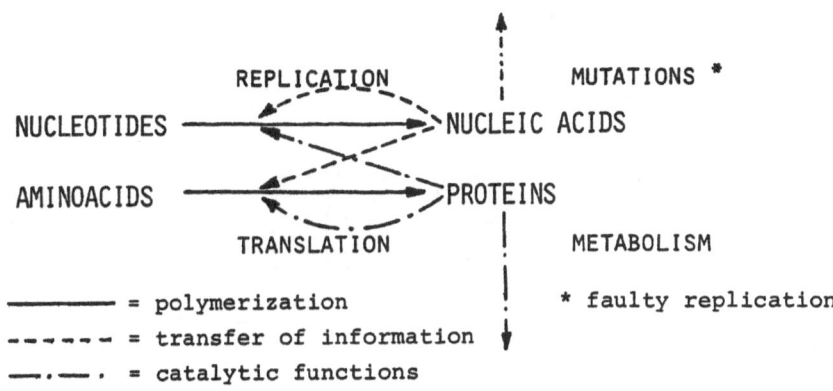

The maintenance of structure may illustrate this scheme. Any living system has a defined structure (which does not necessarily mean a fix form). During all its existence, the basic building blocks of the system (atoms and molecules) are exchanged (metabolism), while the basic structure remains invariant (predetermined information). The nature of this structure is determined by the carriers of information, its maintenance by functional elements (which in some cases may be identical with carriers of information).

Living systems show furthermore a variety of additional properties, some of them specific, some of them common to many systems (organisms):

activity – reactivity – homoeostasis and self regulation – exchange of energy with the environment – self mend and maintenance of structure – adaptation to environment – exchange of information with environment – maintenance of negentropy – behavior modification by learning – program controlled growth (ontogenesis) – consciousness – etc.

Living organisms may, in the sense of General System Theory (Bertalanffy, 1952, 1973; Laszlo, 1973; Weiss, 1969, 1971) be viewed as hierarchically organized, stratified, multi-levelled systems with dynamic interaction between all levels. This may schematically be illustrated in the following figure:

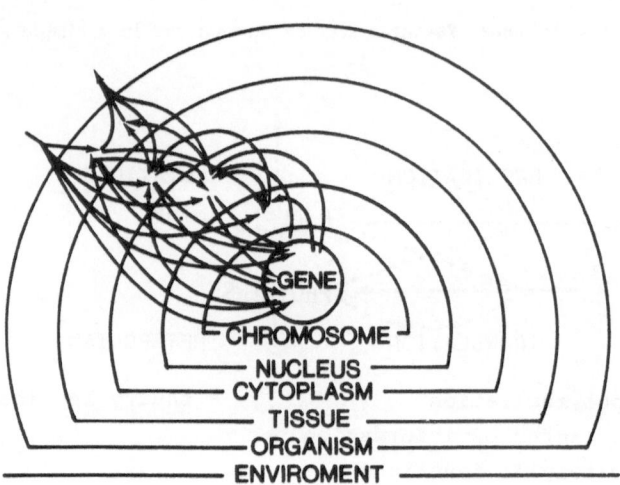

Essentially this means both an interaction between all levels of the hierarchy and a simultaneous superposition of their working principles. But it also means that living systems form **integrated wholes**. An excellent treatment of the topic, both comprehensive and extensive is given by James Greer Miller (1978).

Living Systems (or organisms) are in a sense highly improbable to such an extent that a cynic once defined life as a peculiar disease of matter. As will be shown later, this picture has changed somewhat during recent years – and is still changing. Since living organisms show a very complex structure, the probability of their occurrence by chance is infinitesimally small.

Hence, besides the tendency of matter to organize itself into complex structures, one has to assume the existence of restricting principles which drastically cut down the number of possible combinations of elements; only some of them lead to stable systems. At the atomic level, the Exclusion Principle of Wolfgang Pauli may illustrate this type of restriction (for details see e.g. Gurney, 1934).

During growth and development the complexity of the organism increases steadily. This means that growth implies the occurrence of consecutive states with steadily decreasing probabilities. In order to maintain its structure and life functions intact, the organism has to rely on well—established and stable principles of self—organization, homoeostasis and self—mend. In a certain sense, mechanisms of defence and the ability to change the environment as to fit the needs of the organism may be looked upon as an extension of homoeostatic functions outside the physical boundaries of the organism. Their task is to prevent any disruption of the organism's life functions and integrity.

The concept "life" is intuitively well known, but shows a treacherous elusiveness if one tries to define it strictly. This, of course, is due to the intrinsic complexity of the phenomenon which the term denotes. "Life" is a holistic label which denotes the totality of life functions which characterize a given organism. Furthermore the phenomenon life has several aspects:

a)   the **existence** of a living organism is limited to a **time—span** between "birth and death"

b)   the system properties of an organism **change** during **ontogenesis**, (evolutionary aspect),

c)   the existence of life is dependent on the **state** of the system (living or dead),

d)   the changes which the organism undergos düring its existence reflect the aspects of life as a **process**.

What has been said above furthermore implies that "life" means partly the same, partly different things at different hierarchical levels within the same organism. This could be made clear by an example. Suppose that a living organism, e.g. an animal, is killed. Essential life functions of the organism have been disrupted and it dies. Nevertheless, an **organ**, e.g. the heart may be surgically disconnected and placed into an artificial environment which allows its particular life functions to continue. The organ, a living subsystem of the organism may thus be preserved a long time after the death of the organism to which it once belonged. This was demonstrated nearly 50 years ago (Carrel & Lindbergh, 1938). Now, the organ has a set of life functions, pertinent to its functional role within the organism and may, hence, be destroyed as an organ.

But some of its cells may be kept functionally intact within a suitable environment and may also be stimulated to grow and to multiply. A physical destruction of the cell, however, means a disruption of its functional organization. The cell dies as a living subunit, but still some basic life functions persist: the contractile properties of the actino-myosin molecules, which are responsible for the muscle's ability to contract. This property of actino-myosin (two types of specifically interacting proteins) remains intact even outside the cell-body. If the molecules are dispersed in a suitable solution of nutrients, contraction of the protein aggregate may be triggered by ATP (adenosine triphosphoric acid, a molecule able to deliver energy) (Fulton, 1956). Essentially the same mechanism is utilized when the living cell contracts. The contraction however is **controlled** by mechanisms related to the cell itself or to higher level superstructures. This example illustrates that there is a distinct holistic difference between any system in a "living" and a "dead" state.

But some of its cells may be kept functionally intact within a suitable environment and may also be stimulated to grow and to multiply. A physical destruction of the cell, however, means a disruption of its functional organization. The cell dies as a living subunit, but still some basic life functions persist: the contractile properties of the actino-myosin molecules, which are responsible for the muscle's ability to contract. This property of actino-myosin (two types of specifically interacting proteins) remains intact even outside the cell-body. If the molecules are dispersed in a suitable solution of nutrients, contraction of the protein aggregate may be triggered by ATP (adenosine triphosphoric acid, a molecule able to deliver energy) (Fulton, 1956). Essentially the same mechanism is utilized when the living cell contracts. The contraction however is **controlled** by mechanisms related to the cell itself or to higher level superstructures. This example illustrates that there is a distinct holistic difference between any system in a "living" and a "dead" state.

3. LIFE, COMPLEXITY AND REDUCTION

Basically, living organisms are **very complexly structured chemical systems**. With respect to this, biology favours a reductionistic view. Nevertheless, reduction then leads to the problem of handling complexities. Also, one has to consider the problem of **emerging properties**. In the language of General System Theory, this means that a supersystem may show properties which cannot be predicted solely from the knowledge of its subsystems. Additional knowledge about the functional relationship between the elements of the supersystem would be necessary. Even if it would be possible to derive the general principles which determine a set of functionally equivalent supersystems this would not always be sufficient, since the supersystem may show holistic features which are not reducible.In the terminology of Gestalt Psychology one may say that the whole is more than the sum of its elements. The principle of **superadditivity** represents a still more sophisticated way to describe this fact (Leinfellner, 1984). Leinfellner shows nicely the applicability of this principle to a variety of systems at different levels of complexity. There exists, furthermore, the problem of implicit information, which may be illustrated by an example from genetics. A sequence of codons (triplets of nucleotides from DNA or RNA) determines a corresponding sequence of aminoacids in a protein. But this also implicitly specifies the secondary or tertiary structure of the protein which, in turn, determines its specific functions (e.g. an enzyme function within a given chemical environment). In order to predict this function, one must know the chemical properties of the aminoacids and their particular combinations as well as the distribution of the hydrogen bonds, which determine the higher order structure of the protein; in other words, one has to know the properties of this specific protein.

Formal descriptions of complex systems meet difficulties in defining their different functional levels. While it is easy to recognize an organism or an organ as a whole, there are difficulties with respect to functional subsystems, since there exist multiple (many-many) relationships between functional systems (such as the RAS = reticular activation system) and the anatomical correlates of a system (thalamic nuclei, reticular formation, mediating nerve fibres, glands etc.). The eye, for example, is an integrated anatomical structure but comprises elements which simultaneously are part of an optical subsystem (projection of the image) and an information processing subsystem (visual pattern analysis).

Some of the formal difficulties of defining hierarchical levels in relation to the problem of reduction and emergence are discussed by Bunge (1973) and Kanitscheider (1984). The pluralism of ontological levels reveals itself in terms of "laws" or regularities which are specifically relating to the morphology, stability functions and qualitative features of each level. This fact then is reflected in the different sciences, whose main fields of interest are restricted to a few levels or maybe only to a part of one level. The different sciences are largely interdependent with respect to concepts and methods, but they also utilize a set of methods, concepts and "qualitative descriptions" of their own. Bunge (1982) illuminates this issue in a study about the relationship between chemistry and physics (or properly between quantum chemistry and quantum mechanics).

Any scientific description of our world is reductionistic in the sense that it makes use of the redundancy of the world to describe it simply (Vollmer, 1983). Also, the ability of the human brain to create general concepts means a stepwise compression of detailed information into formally simpler concepts and symbols. pAs Kanitscheider (1984) points out, this type of intrascientific (or perhaps also interscientific) unification is at the level of theories concerned with the phenomenon of surplus meaning: the explanatory or predictive power of the

integrated theory is higher than that of its component theories. One may view this
as an analogy to the earlier mentioned principle of superadditivity.

But already with respect to the description of a system, the symbolic compression
of the description reflects the hierarchical level of the system to which it relates.
This is shown in the following example from genetics: RNA (ion form, pH = 7).

G = guanosine phosphate
C = cytidine phosphate
U = uridine phosphate
A = adenosine phosphate

But there are still problems which relate both to the systemic level and the models we use to describe them. In formalized descriptions, which heavily draw upon structural sciences (such as logic and mathematics) highly condensed descriptions can be gained. An example would be the General Theory of Relativity. But this would normally not be possible in sciences such as geology which rather reflect historical evidence (i.e. boundary and initial conditions). In life sciences of today, the complexity of the chemical systems limits the simplification of their description. While a living system may be analyzed in principle, this may for **economical** or **technical** reasons not always be possible in detail. The structure and genetical layout of some simple viruses such as the bacterial virus O/X 174 (Fiddes, 1977; Sanger et al., 1977) or the Tobacco Mosaic Virus (TM) (Butler & Klug, 1978) are known in detail. The aminoacid array of the TM protein was known already 20 years ago (Stanley & Valens, 1961). The genome of the O/X 174 contains 9 genes of different length and shows a remarkable economy by partial overlapping of genes. The genetic information is stored in a circular DNA molecule of 1.8 μm length, comprising 5,375 nucleotides. A print-out of this genetic array (using letter abbreviations) would demand one A4 page. A corresponding description of our intestinal bacterium, Escherichia coli, would probably demand a space equivalent to several volumes of Encyclopaedia Britannica. Finally then the implicit information regarding the proteins would demand some additional volumes.

The term "reduction" here has intentionally been used in a loose way, mainly in the sense of partial reduction. A more thorough discussion of the concept is given by Vollmer (1984). The essential point is that living systems cannot be described in terms of a single science. An understanding of the working principles of any life function demands a partial "reduction" of its description to the conceptual framework of more "basic" sciences. Suppose, one wants to study the chain of events which lead to visual perception. Visual information is mediated by light, a **physical** phenomenon. Photons are absorbed by photopigment molecules(e.g. rhodopsin) of retinal receptors. This results in a photochemical excitation of the molecule, a **quantumchemical** event. Rhodopsin is dissociated into a very specific protein, opsin. and the aldehyde retinal, which simultaneously undergos transformation in that the light-sensitive cis-retinal is switched into the insensitive trans-retinal. The photopigment molecules are contained in small circular compartments, which are enclosed by a plasmamembrane. Together they form, like a pile of hollow coins, the main body of retinal rods (**descriptive microanatomy**). The light activated rohodopsin molecule has a high energy content. By a complex chain of chemical events, an electrical current of Na^+-ions, which normally passes through the plasma membranes, is changed. This results in an electric generator-potential at the synoptic end of the receptor cell (**electrochemistry, electrophysiology**). Finally, by spatio-temporal summation effects from several contributing receptor cells, a nerve impulse is triggered in a ganglion cell of the optic nerve (**electrophysiology**). Nerve impulses from several neurons are then redistributed at a thalamic relay station in the lateral geniculate bodies. The sequence of nerve signals is then analyzed in the primary and secondary "centres" of the occipital visual cortex (area striata) with respect to their information content. The latter is related both to the pattern of nerve impulses and to the channels through which they arrive. The last mentioned principle may be viewed as a "principle of specific termination", a modern re-formulation of Johannes Müller's famous doctrine of "specific sense energies" from 1826 (**neurophysiology**). Finally, the neurophysiological state of the brain is changed in a specific way and accompanied by the conscious perception of "light" (**psychophysiology**).

This very incomplete description, nevertheless, illustrates the above mentioned thesis about interscientific relationships in any description of living systems. (For further details see: Gemme & Bernhard, 1975; Lindsay & Norman, 1973; Zurer, 1983).

There are aspects of life which could not be understood in terms of the models of classical science. The multidimensionality of its conceptual background and the complexity of the systems to which it relates is only part of it. Nevertheless, these difficulties promoted the emergence of vitalistic theories and supported their existence up until today. New varieties of vitalistic models still arise today. An example is Rupert Sheldrake's hypothesis about "invisible morphogenetic fields".

Similar ideas have, however, been expressed earlier by the German biologist and philosopher Hans Driesch. Vitalistic theories try to relate the existence of life to the effects of a non-physical "life force". Additional conceptual difficulties arose from the introduction of divine or other spiritual principles in connection with teleological speculations about the "purpose" or "meaning" of life within the "great scheme of our universe". A good survey of these aspects is given by Wuketits (1982); see also chapter 1 of this volume. It is important to remind oneself of the influence which the "Zeitgeist" or certain ideas may have on the impediment or promotion of scientific progress. At the time of Lavoisier (late 18th century), "organic" compounds were thought to be dependent on living organisms for their existence (the notion "organic" for carbon compounds still reminds us of this fact). In 1828, Friedrich Wöhler (a pupil of Justus Liebig) succeeded in synthesizing urea from ammoniumcyanate:

$$NH_4CNO \longrightarrow (NH_2)_2CO$$

ammoniumcyanate   *urea = carbamide*

The discovery was important not only to chemistry, but it questioned vitalistic theory, since urea is typical metabolite from animals and ammonium cyanate could be synthesized directly from its elements. This example illustrates nicely how our opinions about which questions principally can be answered (in this case the "reduction" of organic matter to inorganic) is dependent on the actual level of knowledge. This and similar breakthroughs now stimulated the rapid development of biological disciplines into science in a modern sense, which employs hypotheses, critical tests and predictive theories. The same then also applies to medicine as a field of application of biological knowledge. As we know today, some main steps in this direction were the development of genetics, cellular theory, the biochemistry of metabolism and the application of concepts and principles from physics to living organisms.

4. LIFE AND ENTROPY

One principle of the above mentioned type is entropy. For many years a too narrow interpretation of classical thermodynamics has been an obstacle for an understanding of living systems. One did not always realize the fact that the Clausius principle was formulated for closed systems. In our own time, a misinterpretation of thermodynamics in this respect is used as an argument by adheres of the "Scientific Creationist" movement which denies Darwinian evolution (Pierce, 1981). They do not realize, however, that a validity of their arguments would make ontogenetic evolution impossible as well. An analysis of this type of arguments is given by Radner (1982). From an epistemological point of view, Radnitzky (1983) shows, that models of this type do not satisfy necessary criteria of scientific theories.

Living organisms are thermodynamically open flow systems, which exchange matter and energy with their environment. This may schematically be illustrated in the following figure:

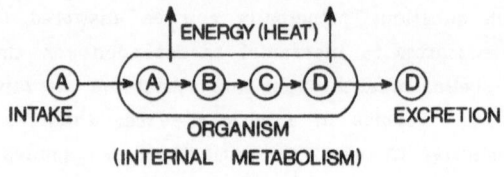

A , B , C , D = chemical compounds (metabolites)

Generally, the following energy relation holds: $E_{(A)}$ > $E_{(D)}$ (E= energy of compound) A candle light is an example of a simple flow system.

With respect to entropy, an organism may be viewed in relation to its environment. A change of entropy can be split up into at least two parts, an external one (e) and an internal one (i):

   dS = (dS$_i$ + dS$_e$) ≥ 0 (dS = 0 in a stationary state)

where dS$_i$ < 0 may be possible, as long as dS ≥ 0, which is the case if |dS$_e$| = |dS$_i$|

Flow systems tend to decrease their entropy when moving from an arbitrary initial state to a stationary final state, which represents a stabilized equilibrium (equifinality in dissipative systems). A strict treatment if the subject is given by Prigogine (1967, 1976); Nicolis & Prigogine (1977).

In this way, the living organism during its life keeps a high level of negentropy (low entropy) which is made possible by the use of energy rich components from the environment and by the expenditure of an excess of energy. Kenneth Sayre describes this fact in the following way: "Life thus, is a catalyst working to increase disorder in the hosts environment, acting more efficiently in life forms that are more complex" (Sayre, 1976, p.93).

The last statement then leads to a final issue: the relationship of entropy to order. This aspect of entropy appeared first in statistical mechanics. Energy states of high entropy have high probability. So have systems with high degree of randomness (lack of order). The thermodynamical entropy (S) of a physical system (of molecules) is according to Boltzmann a logarithmic function of its thermodynamic probability (P) of its occurrence:

$$S = k \cdot \ln P \qquad *)$$

(k = Boltzmann' s constant = 1,38 . 10^{-23} J/deg; for details see Fermi, 1956).

In a similar way the concept entropy appears within the Mathematical Theory of Communication. The information content (H) of a message (information theoretical entropy) on the other hand, is a negative logarithmic function of the probability of its elements:

$$H (A) = - \Sigma P_a \qquad {}^2\log P_a \qquad *)$$

(a are the elements of an ensemble A (a message); if binary logarithms (^{2}log, base 2) are used, H is expressed in bits).

Thus both types of entropy and the concept of information are related to the probability of a state of a system. The main formal difference is the minus sign. The relationship between the concepts has been interpreted so, that information theoretical entropy (information content) can be looked upon as a negative function of thermodynamical entropy, which means that H increases as S decreases. This fact gave rise to Brillouins and Wieners characterization of information as negative entropy or negentropy. (Wiener, 1961, Brillouin, 1962). The deeper implications of this relationship have, however, been the object of different interpretations.

The transfer of information means the transfer of structural of organizational principles. Living organisms receive their vast store of basic information through genetic instructions, which is stored and encoded within the DNA of genes. Later on the organism receives information through sensory processes and stores it in different types of memories.

*) Strictly speaking, P is not a probability in a statistical sense, but a large number, denoting the amount of possible microstates of a system that correspond to a given macrostate. However, the thermodynamical probability is proportional to the statistical.

*) H is a general concept which may be related to any probability density function $0/(x)$ of a stochastic variable x:

$$H = - \quad 0(x) \; \log \; 0(x) \; dx$$

A short note concerning information and negentropy should be added. One has to
be careful how to use the terms. The transfer of information proper in the sense
of Shannon's Theory is formulated for energetically closed systems. The theory
presupposes the existence of established channels, transmitters and receivers. No
net energy is transferred to the receiving system, only the structuring impulse
(information). Meanwhile, all living systems are thermodynamically open systems,
exchanging energy and matter. The apparent contradiction may be solved by
assuming that information is gained by different processes at different stages of
ontogenetic development and by different processes at different levels of the
hierarchy.

*   The increasing order (negentropy) of organisms during ontogenetic growth is
    due to the effect of genetically transferred instructions (order through order).

*   Transfer of information in the sense of Shannon's theory starts in connection
    with learning and feedback processes when underlying mechanisms have been
    developed. This type of information transfer is restricted to certain
    structures, which at the given hierarchical level of the system behave as
    thermodynamically closed subsystems. Hence, this presupposes the existence of
    established networks (subsystems) which encode transferred information in a
    way which confines it to the boundaries of this subsystem. It also implies
    that the internally coded information is comparatively stable to the influence
    of metabolic processes, which always bring about a change of thermodynamical
    entropy.

5. PHYLOGENETIC ASPECTS OF LIFE

It has been mentioned earlier that the emergence of living organisms – as we
know them on earth – is bound to the existence of given properties of the
universe as necessary preconditions but also to the existence of favourable local
conditions. Only a few of them could be mentioned here:

*   A planet with suitable mass and atmosphere.

*   An optimal distance from a sun of suitable spectral class. These factors are
    partly interdependent.

*   The availability of certain elements in sufficient quantities (C, H, O, N etc.)

*   The homeopolar bond of carbon atoms, ensuring the formation of large carbon
    chains.

*   The surprisingly high amount of "abnormal" properties of common water; two
     of them are: a density maximum at 277 K (3.9 degrees above the freezing
     point); the ability to work as a nearly universal solvent, comprising both
     salts and non-polar compounds such as sugars. These properties depend on
     the pronounced dipole moment of the water molecule, a factor which then
     largely influences the geophysical and geochemical background on which the
     existence of our biospehre depends.

Undoubtedly chance events must have influenced the emergence of life on earth.
During the last 15 years however, new models and discoveries have modified this
view. Some of these models have emerged within physics and chemistry. One
approach is that of Prigogine and co-workers, which has shown that quite new
properties can emerge in systems which are far from equilibrium in a
thermodynamical sense. Dynamically unstable systems of this type gain a new type
of order through fluctuations. Dissipative systems are examples of systems in
which new structures emerge, which stabilize the system with respect to
fluctuations. This means a kind of self-organization.

Mathematically, these systems are described by non-linear equations. Their
importance was realized after the introduction of the computer which made a
numerical treatment of non-linear terms feasible. Due to computational difficulties
they were earlier sometimes neglects as "second order" factors. It is known today
that non-equilibrium dynamical systems may be stabilized by their non-linear
properties. Also general evolution criteria have been derived (Glansdorff &
Prigogine, 1971).

Models with similar features arose from the study of Laser action (LASER = Light
Amplification by Stimulated Emission of Radiation). If a laser device is excitated
by very strong energy currents (e.g. light) above a certain critical value, the
atoms of the device (e.g. "doped" crystals) show a spontaneous cooperative
behavior: a high intensity beam of coherent light is emitted. The (rather
complicated) general principles behind cooperative systems have been formalized by
Hermann Haken under the heading of synergetics (Haken, 1976). An interesting
synopsis of spontaneously ordered systems in an essentially non-mathematical
language has been given by Sexl (1984; see also Schopper, 1984). Models of this
type seem applicable to a variety of systems and phenomena from RNA, DNA and
proto-life to cooperative brain functions and the spread of ideological prejudices.

Finally, Eigen, Schuster, Winkler and others have shown the applicability of
cooperative game theory (as part of a theory of dynamic differential games) to the
"struggle" of DNA molecules for survival by cooperation with proteins (Eigen, 1971;

Eigen & Winkler, 1975; Eigen & Schuster, 1977, 1978). Implications of these models
have been analyzed by Leinfellner 1983, 1984a).

Our knowledge about the origin of life on earth is, to a large extent, based on
conjecture, even if some details are known with reasonable degree of certainty
(detailed references are given by Löwenhard, 1981). The age of our earth is
estimated to about 4,600 billion years. The first signs of life seem to have
appeared some 3,200 to 3,900 billion years ago (Halstead, 1975, Sagan, 1978). The
timepoint somewhat depends on where (at which level of complexity) the limit for
forms of life is set. With a high degree of certainty, only simple carbon compounds
(chains of 2 to 3 atoms) existed 4,500 billion years ago, while 3000 to 4000
billion years later highly differentiated life forms had emerged (Ehrensvärd, 1961).
The oldest known microfossiles dated back to about 3,500 million years ago
(Mustelin, 1983). Microscopic fossils, identified as remnants of the earliest
eukaryotes (originally called acritarchs = of uncertain origin), have been shown to
be unicellular planktons, about 1,400 million years old (Vidal, 1984). Recently it
was mentioned (Lagerkvist, 1983), that prokaryotes and eukaryotes might have a
common origin in archaebacteriae.

One can arrange systems in order of increasing complexity. The given scale (see
figure) should be interpreted as an ordinal scale, since no estimates of the real
magnitude of complexity can be given. This presentation does not, of course,
consider qualitative differences between life forms of approximately the same level
of complexity, which would imply a comparison between plants and animals. The
increase of complexity reflects the emergence of additional life functions. Proteins
and nucleic acids show some of the basic properties which are the foundation of
life functions proper. Aggregates of complex molecules show an increasing
versatility in their reactions with the chemical environment. With increasing
complexity the systems gain independence from their environment. This means that
their demand on the environment with respect to certain premanufactured
components decreases. At the same time, more regulative functions are incorporated
within the boundaries of the system. Protolife thus may have existed in a variety
of forms. RNA and DNA seem to be the first steps to self-replicating molecular
systems. Leinfellner (1984) cites a discovery of Cech (1984) according to which a
single stranded RNA (occurring within a host organism, Tetrahyma) is able to
reproduce itself without the help of proteins, i.e. "partogenetically". However, as
Eigen and Schuster have proposed, a cooperation of DNA and proteins show
superior qualities in the formation of "gene life". The next step then is the
formation of holo-life organisms which incorporate all necessary life functions into
their boundaries.

The emergence of early forms of life and their development into the existing forms, reflect a complex interaction with an interdependence on that environment which gave birth to them. According to Haldane and Oparin, the early predominantly reducing atmosphere (anoxygenic atmosphere) of earth was supposed to promote the origination of a chemical background from which the earliest protoforms of living systems could emerge (Dickerson, 1978; Mayr, 1978). An important mark of progress is the emergence of closed cells, protected from environment by membranes which facilitate the selective absorption of nutrients and the excretion of metabolites. The earliest types of cells, prokaryotes, lack nuclei. They are represented by bacteria and dyanobacteria. The next essential step is the development of eukaryotes, cellular organisms with distinct nuclei. All higher plants, animals and fungi are based on this type of cell.

For several billion years, one-celled mainly anaerobic organisms were the only forms of life. However, these early primitive microorganisms gave rise to biochemical systems and the oxygen enriched atmosphere on which modern life depends (Schopf, 1978).It also seems that these changes created the necessary conditions for a more rapid diversification of life forms and the emergence of multicellular organisms with steadily increasing efficiency.

After the first stable forms of life had been established, further evolution was mainly governed by two mechanisms: genetic variation and selection. Genetic variation is accomplished by mutations, i.e. rearrangements of molecular sequences in DNA and RNA. Another mechanism depends on the exchange of corresponding chromosomal segments between homologous chromosomes by formation of chiasms and crossing over. This means a rearrangement of genetic information at the chromosomal level. Gene drift and jumping genes are mechanisms which were proposed more recently.

Genetic variation is a necessary precondition for evolution. Modern molecular genetics have also shown that variations within a species are much larger than Darwin once postulated (Ayala, 1978). The present view is confirmed by studies about protective colour adaptation for different kinds of insects in relation to a changing environment. While random mutations safeguard continued variation and recombination for the development of new capacities to deal with environment, the test of survival accomplishes the natural selection of organisms which are best adapted to existing conditions.

The importance of "system theoretical" aspects in relation to phylogenetic evolution may be summarized in the words of Dobzhansky: "Adaptation and emergence of new genotypes is a feedback process within a reproductive group, natural selection is homoeostasis within a relatively isolated biotic system. The results of these two processes operating together is the evolution of established species" (Dobzhansky, 1955, p. 131).

Cybernetic self-regulation is a universal system property. While homoeostasis may be defined as **adaptive self-stabilization**, **adaptive self-organization** characterizes selective progression towards the emergence of better adapted species, capable of handling increasing amounts of environmental changes:

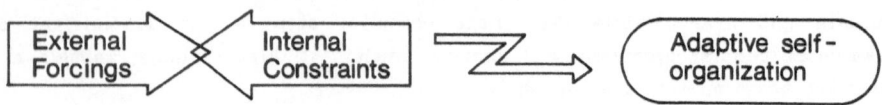

A mark of success of an organic form in evolution is the ability to achieve stability within their environment. This amounts to the persisting ability on the part of the organism to feed upon the negentropy of its immediate environment (Schrödinger, 1951). Sayre coins, in this connection, the term negentropic flexibility: "In speaking of flexibility in the assimilation of negentropy I refer to the capacity of an organism to establish efficient couplings within its environment, under a range of different conditions, through which negentropy can be obtained to support growth and metabolism and to control its respondenses to environmental contingencies. Let us name this capacity negentropic flexibility. Mobility supports the development of negentropic flexibility. Individuals with superior mobility will tend to replace others within their reproductive group, thereby strengthening the genetic factor that makes mobility possible. Among such factors, however, is perceptual sensitivity, which contributes directly to the organisms capacity to acquire negentropy in form of information". (Sayre, 1976, p.117).

If one looks for the results of evolution, reflected in life forms of today, one finds a large variety of interdependent living organisms, where complex multicellular organisms coexist with unicellular ones and with protoforms of life. It is interesting, and from the standpoint of the scientist a lucky circumstance, to find a reflection of phylogenetic evolution in the existence of a whole scale of organisms with varying degree of complexity. All life forms together, fauna and flora, constitute the large ecological system of life on earth. This does not imply that simple life forms such as bacteria of today are identical with their early ancestors. Their total amount of information may approximately be unchanged, but the "quality" of their genetic information, expressed in terms of the organism's biochemical efficiency, is probably higher in organisms of today.

The purpose of this chapter has been to show that "life" from a phylogenetic point of view is no unitary concept. Rather the meaning of the term and its range of connotation changes with the increase of complexity of the systems (organisms) to which it is applied. There seems to be no definite demarcation which allows an exclusive classification into so-called living and non-living systems. Rather there is a continuous transition from non-living matter which nevertheless may show typical basic life functions, via an extended range of uncertainty to definite "living" systems. The transitory types of systems show the gradual incorporation of an increasing number of homoeostatic life functions and metabolic processes into their boundaries.

6. THE ORGANISM'S INFORMATION SYSTEM

The earlier paragraphs have dealt with some of the main features of living organisms. As far as we know, these are the necessary preconditions for the emergence of awareness or consciousness. The emphasis will now be shifted towards the specific structures to which consciousness is related:

the latter are information systems. Formal, a distinction must be made between functional systems such as the CNS and their anatomical correlates, such as the brain. While for reasons of convenience a mixed language is used in this paper, it is essential to remind oneself of the fact that there exists a multiple relationship between functional subsystems and anatomical structures. Otherwise, one easily ends up in phrenology.

The following paragraphs will deal with functional properties of the brain and CNS. Of course, only a few essential features can be considered. Neglecting all details, the information system of higher animals may be summarized in the following diagram:

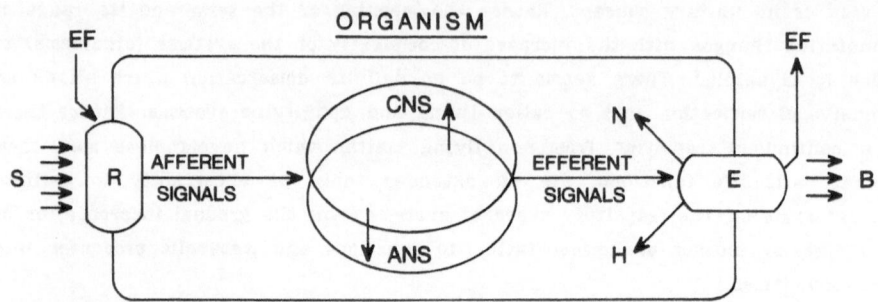

R = RECEPTOR UNITS   E = EFFEKTOR UNITS (MUSCLES & GLANDS)
S = INFORMATION INPUT BY STIMULI
B = INFORMATION OUTPUT BY RESPONSES (BEHAVIOR)
N = INTERNAL NEURAL FEEDBACK   H = INTERNAL HORMONAL FEEDBACK
EF = EXTERNAL SENSORY FEEDBACK (VISUAL,AUDITORY INFORMATION)

Any physical event which affects the organisms sensory equipment is called a stimulus. While stimuli differ greatly with respect to their physical nature, their common function is to convey information from the external world or from the organism itself. Any transfer of information means a transfer of organisational or structural principles which increase the order of the receiving system. With respect to the dynamic activities of the system, a transfer of information decreases its uncertainty in choice situations. This applies to all levels of the systemic hierarchy. The function of receptors is to mediate the transformation of physical stimuli into nerve signals. The essential aspect of this process is a re-coding of the information content of the stimulus into a different neural code. In this way, a consecutive information processing is performed. While the main information content (the essential message) of the stimulus remains invariant, it gradually becomes part of a larger pattern. Two principles of coding seems to be available: the formation of spatio-temporal patterns of nerve impulses and the principle of specific termination, i.e. the relationship between certain types of information and the channels through which they arrive. This was mentioned earlier in connection with an example on visual information processing. It should be added here, that "specific termination" may not be interpreted as a single terminus, but as a set of neurons, distributed over large parts of the cortex. Meanwhile the signal which pass down a certain channel also pass specific relay stations as well as cortical areas which always participate in the processing of a particular type of information. As an example it may be mentioned that pitch discrimination of sounds mainly, but not solely, is related to a utilizing of specific channels (or more precisely to signals which originate from specific frequency sensitive loci of the inner ears basilar membrane), while loudness discrimination is mainly based on the impulse density within volleys of nerve impulses.

As to the distribution of terminals, the essential features of such mechanisms were on theoretical reasons recognized more than 30 years ago by Friedrich von Hayek, who (as a "non-professional" psychologist) gave an outstanding analysis of the problems of perception (v. Hayek, 1952). We will return to this question later on.

The animal organism uses two types of information systems in parallel: the nervous system, which starts rapidly and the endocrine (hormonal) system, which starts slowly, but has a more enduring action. The former system is based on a mediating of electrochemical signals (action potentials) in conducting fibres (neurons) and may essentially be compared with telegraph lines. The endocrine system utilizes signal molecules (hormones), which are injected into the blood stream and in this way distributed all over the organism. They affect, however, only specific receptors, which chemically respond to a certain hormone. Receptor sites may be located at glands (which allows for hormonal feedback mechanisms), at other organs, such as the heart, or at the synoptic junctions between neurons. The transfer of nerve signals from a presynaptic to a postsynaptic neuron is mediated by neurotransmitter molecules which, in some cases, are identical with hormones. In this way the endocrine and nervous systems interact at the neural synapses. This implies that a certain message during some short time of its existence is encoded in terms of signal molecules. It is also the reason why a change of the biochemical environment of the synoptic system may change the information content of the message which passes the system. The action of psychotropic drugs may be viewed in this way. Their medical use is aiming at a re-normalization of a distorted message. The details of these processes, while at least partially known today, are too complicated to be dealt with in this paper.

An understanding of the working principles of our brain has in many ways been facilitated by computer technology, which in turn has profited from brain research. While there are similarities between brains and computers, there are also essential differences. Basically both are information processing systems. The computer, however, is designed and presupposes the existence of brains. The latter are the result of an essentially "self-sustained" evolutionary process. The computer depends essentially on deterministic principles, components of high reliability and the use of rapid sequences of electromagnetic pulses. The brain uses partly stochastic principles and is rather insensitive to the failure of single elements (neurons). As a chemical system it depends on rather slow mechanisms of ion transport through membranes, but uses many lines in parallel, has a high degree of economic flexibility and performs a sophisticated processing of information at the synoptic relay stations.

The following description may illustrate some of the synoptic functions with respect to quantitative information. The eye receives every second in the order of 10^8 bits of information, while only about 50 bits per second are recorded at the highest levels. What happens to the rest? Suppose, that according to the none-or-all principle each nerve impulse corresponds to one bit. Normally a certain number of nerve impulses summarize their effects at a synoptic junction in order to trigger a postsynaptic nerve impulse. This includes a superposition of excitatory and inhibitory actions.

Suppose that an equivalent of N excitatory input signals is needed in order to trigger a postsynaptic neuron within a critical time period Δt (the effects of input signals are of short duration). Now, a large number N decreases the probability that a postsynaptic neuron is triggered by spurious random signals. Drugs or hormones may change the threshold value of the synapse i.e. the number N (alternatively the time Δt). This means that the synapse works as a coincidence detector with independently controlled time constant. The synapse reduces the number of bits which pass the synapse by a factor (1/N). This "degradation" of information means that the system pays with information in order to gain certainty about whether something at the output really corresponds to something at the input (McCulloch, 1951). This principle then reflects some of the consequences of the hierarchical order of living organisms with respect to information processing. A large amount of "low grade" information at the input emerges as a small amount of "high grade" information at the highest levels.

Nevertheless, the reverse principle coexists. A single train of nerve impulses may originate from a pain receptor. After having passed several relay stations, the signals at the spinal level already engage five different neural pathways. The final brain state includes contributions from several cortical and subcortical structures. The corresponding perception of pain then represents complex information about e.g. the site and type of injury, the intensity of pain, its emotional aspect in relation to a knowledge about the danger of the injury, the attitude of the person towards pain within a social context, etc. (Melzack, 1961). Further aspect of brain functions will be dealt with later on.

Essentially the brain may be viewed as an autoanalytical instrument which is able to detect and record its own states. While the brain is endowed with the superior principle of conscious information processing, some autoanalytical functions may already be accomplished by a suitable hierarchy of analyzing programs, where higher level programs evaluate and correct lower level programs with respect to external criteria of efficiency. In a sense, the brain may be viewed as a universal Touring machine (i.e. a machine which is able to simulate the functions of any other machine). The programs may then be viewed as virtual machines, which govern the creation of physical machines (tools, engines, computers) by direct control of manipulatory equipments (arms, legs, feet, hands). The really marvellous accomplishment, however, is the brain's ability to create its own programs.

The above mentioned autoanalytical functions means that the brain has access to two different sources of information: external and internal ones; both however, are integrated in an internal representation of the organism itself.

The availability of two distinctly different types of information, representing different aspects of reality, historically gave rise to the so-called psycho-physical problem, an experimental reformulation of the body-mind problem. In a modern language, it may be said that neurophysicological events and mental (cognitive, subjective) events are different aspects of the same brain functions. Methodologically, these aspects are revealed in two sets of data which are correlates of each other.

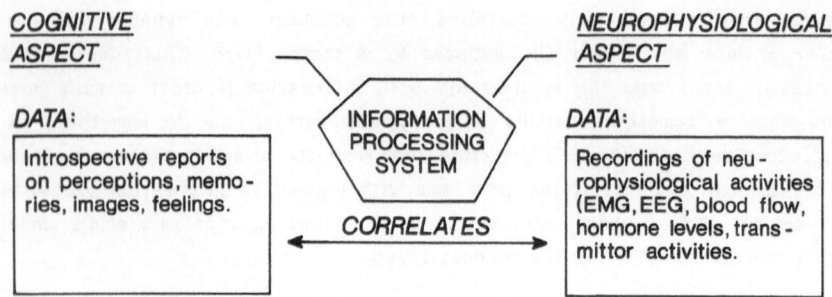

A simple example may illustrate this. If the eyes are stimulated by periodic flashes of light at a low frequency, each flash is perceived separately. A simultaneous recording of the EEG (electroencephalogram) from the occipital lobes of the cerebral cortex shows a typical pattern of so-called evoked potentials:

Notwithstanding some short time delay, the experience of light flashes and the electrophysiological potentials correspond to each other; they are correlates. If now the flicker frequency reaches a certain critical value, CFF (=critical flicker fusion frequency), the flashing light is perceived as a steady one. At the same time, the distinct evoked potentials disappear from the EEG. The CFF-value is a logarithmic function of light intensity:

$$CFF = a \cdot \log I + b$$

(a and b are constants). The CFF phenomenon makes it possible to use alternating currents (50 – 60 Hz) for illumination purposes.

The final issue behind the psycho-physical problem, nevertheless, is the phenomenon which we call consciousness.

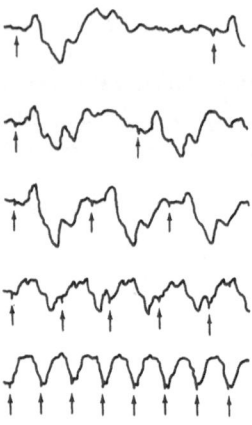

7. CONSCIOUSNESS

The previous chapters have dealt with the conceptual and theoretical background of the model which is given below. Since "life" and "consciousness" are somewhat ambiguous terms, the terminology used should be clarified. A more detailed discussion is given by Löwenhard (1981).

Consciousness is used as a technical term which denotes what in a subjective language is described as awareness or subjective experience.

Awareness, hence is used as an equivalent to consciousness from the standpoint of the subject.

Sensation is used as a general term to denote the effects in terms of subjective experience of the activation of receptors. A clear distinction is made between the phenomenon of consciousness as such and mental objects (symbols, images) which represent the information content of a conscious brain process.

Feelings is an ambiguous term which is used to denote unspecified qualities of awareness, which are neither related to mental objects nor specific emotions, acts of will or motivational states (hunger, thirst etc.). One should note, however, that "feelings" in everyday language includes moods, emotional and motivational states as well as sensations such as pain (see Jaspers, 1948; Zethraeus, 1962).

Self–consciousness is not used as identical with consciousness, which has a larger connotation. The distinction is treated later on.

Mind is used as a label to denote the "non material universe of mental phenomena", i.e. "world 2" in the terminology of Karl Popper (Popper & Eccles, 1977). It is the conceptual framework to which all phenomena of conscious experience are related. The term does not imply any connotation of a homunculus (see Crick, 1979). The following statements may be regarded as speculative hypotheses, or, in some cases as descriptions of observable phenomena. While the underlying model essentially corresponds to evolutionary epistemology and the phenomenon of consciousness is viewed in analogy to life, there is a difference: Life can, at least principally, be explained in terms of known principles, even if many details still are obscure. Contrary to this, consciousness still holds features which presently are outside the realm of a strict scientific treatment. This does not mean they will be so in the future.

I. Consciousness, as we know it, seems to be restricted to a certain class of living organisms, utilizing nervous systems (brains). The human brain, and probably that of other animals, seem to use both "conscious" and "non-conscious" modes of information processing. Consciousness may be described as a phenomenon which accompanies certain modes of information processing, characterized by direction of attention onto the information content of the on-going process. This amounts to an ability by the participating nerve net to sense its own states. Consciousness is, hence, a principle which strongly enhances the brain's ability to work as an autoanalytical instrument. It should be noted, however, that the subjective nature of "immediate knowledge", which characterizes conscious information processing is no **necessary** condition for a complex system to "sense" (= detect and record) its own state as long as one allows for a suitable hierarchical organization of programs. Nevertheless, it seems that the principle of consciousness allows to circumvent this condition, which means that certain nerve nets are able to perform the task of sensing their states within their **own** functional level. This reflects the peculiar self-referring or reflexive properties of such systems. In this connection it is essential to make a distinction between the fact of "being aware", i.e. the participation in an immediate act of experience, and knowledge about the principles which underlie the emergence of consciousness. This knowledge, nevertheless, may be the symbolic content of a conscious nerve process.

II. Consciousness seems to arise due to system properties which a computer lacks. It is probably not lonely a consequence of a system's complexity as such, but reflects inherent properties of neurons to which the phenomenon is related. The critical system parameters are, as yet, not known. The essential underlying properties may be related to the cellular level, the subcellular level or both, but consciousness seems to manifest itself only clearly in sufficiently complex systems. Nevertheless, the essential genetic instructions necessary to develop the system properties which underlie the emergence of consciousness, have to be encoded into the genome as part of the explicit and implicit genetic information. Some speculative assumptions may be made:

1. Consciousness, like life, are basic attributes of our universe, but manifest themselves only under specific conditions in complex systems.

2. One underlying basic property may arise from an extrapolation of the excitability of complex aggregates of protein molecules.

3. Consciousness is a phenomenon, which has an extension in time. It may be dependent on spatio-temporal patterns of slow potentials and discrete pulses (bioelectric fields) which are related to aggregates of nerve cells and which fulfil certain, yet unknown conditions regarding structure and temporal continuity ("world lines"). A model of this kind which introduces the last mentioned features under the heading of "historical causality", is given by Culbertson (1963, 1982). As to neurophysiological data, John (1968) has in

connection with mechanisms of memory shown the transformation of local potentials into patterns of nerve impulses and their reshaping into similar potentials at other parts of the cortex. Essentially, this amounts to an analog–digital–analog conversion.

4.   In analogy to life, consciousness is **state–dependent**. It is empirically known that a number of conditions must be met in order to let a conscious brain state occur. Such conditions are the activation of the cortex by the thalamo-reticular system (ARAS), an on–going electrophysiological activity (EEG) within the cortex and an undisturbed blood flow, ensuring the availability of oxygen and nutrients on which the nerve cells depend.

It may be added that the particularly vivid experiences during sensory perception seem to demand both cortical **arousal** which may be defined in terms of phasic physiological responses to sensory input, and **activation** defined as tonic physiological resdiness to respond (Pribram & McGuiness, 1975).

III.   Consciousness seems to change with respect to both quantitative and qualitative aspects as a function of the complexity of the nerve net to which it is related. Thus, the concept of "consciousness" is, in analogy to "life" not a unitary concept.In a phylogenetic sense consciousness seems to develop alongside the new system properties of living organisms. It may be assumed that consciousness increases both with respect to quality and scope from primitive awareness at lower phylogenetic levels to abstract thinking and self–consciousness in humans alongside an increasing complexity of the corresponding brain structures. While, of course, no one can prove that for example the bee has awareness, the possibility of it cannot be denied, even if awareness her would mean a rather rudimentary form. One may, of course restrict the use of the term consciousness to the human level, which essentially means that one makes it interdependent on verbal communication (see e.g. Eccles, 1970), but this would not answer the main question, since awareness seems to be prior to language ability. Furthermore, one would have to decide at which level of complexity and according to criteria (or phylogenetic level) the limit for the existence of consciousness is set. There are strong reasons to believe the existence in most animals of sensory awareness as well as different types of cognitive abilities. Elementary mental functions have been shown to exist in very simple systems of neurons and isolated neurons (Kandel, 1979; Sokolov, 1981; Sokolov & Grechenko, 1981; Sokolov & Willows, 1981). Also there are several reports which indicate dream activity in animals (Sagan, 1978). However, a cat is not a dog and a chimpanzee is not a human being. One should be careful not to project **human** concepts or modes of experience into animals. This is part of the relevant criticism which early behaviorists made against the animal psychology of the 19th century. There is reason to believe that the perception of the external world shows species bound variations which relate to the quality of sensory mechanisms as well as to their adaptation to a certain ecological niche. As long ago as 1928 von Uexküll stated: "It would be a very naive sort of dogmatism to assume that there exists an absolute reality of thins which are the same for all living beings. Reality is not a homogeneous thing, having as many different schemes and patterns as there are different organisms". From what is known today, it is probably not possible to state the existence of a definite borderline. Rather, in analogy to "life", it seems probably that there is a transition from "unconscious" life forms via a range of uncertainty to definitely "conscious" ones. Due to the lack of space, further details have to be omitted (but see Löwenhard, 1981, 1982a). From a much broader point of view, the evolution of the human mind and its cultural products has been the object of evolutionary epistemology. Only a few references will be given here: Bartley (1982), Campbell (1974), Lorenz & Wuketits (1983), Löwenhard (1982b), Radnitzky (1983), Riedl (1979), Vollmer (1981,1982), Wuketits (1982b).

IV. The brain uses both conscious and non-conscious modes of information processing. Strictly speaking, one has to make a distinction between **unconscious** processes (information not retrievable except by inference), **subconscious** processes (information only retrievable under appropriate circumstances) and **non-conscious** processes (information not stored in memory); (Hilgard, 1980). Normally, different modes of information processing are used simultaneously. The activity of the ANS is non-conscious, but the brain may be aware of its **effects**. Our conscious experience can be characterized by **content, modes of consciousness, sense modalities** and **states of consciousness**. It is probable that both different modes and different states of consciousness relate to either different **processes** or different **brain states**. Different modes of consciousness seem to be related to different stages of information processing. This includes a reactivation of memory content, images etc. A model of this import has been proposed by Aurell (1979, 1983). It is well known, that the vivid intensity of sensory awareness is different from the pale reproductions of past experiences. Nevertheless, there exist conditions during which images may reach some of the qualities of actual perceptions: hallucinatory states which may be induced by hypnosis, intoxications, infections or mental diseases.

Subjectively, we speak of different **states** of consciousness which correspond to different neurophysiological states. It is clear, however, from what has been said above that there are brain states which do **not** correspond to any conscious experience. One generally makes a distinction between normal states of consciousness (NSC) and altered states of consciousness (ASC); (see e.g. Krech et al., 1982). NSC vary with vigilance (effects of arousal and activation), attention (selective cortical activation) or changes in alertness due to circadian rhythms. ASC occur naturally during sleep and its different phases, during transitory states, such as the hypnagogic state (= transition from NSC to sleep) or as an effect of psychotropic drugs.

Our sense modalities are known from experience. It is a different sensation to hear than to see. The implications of the concept are clear from the fact that sense modalities represent **distinct** internal sensations, which means that a given modality cannot be transformed to another. It is normally not possible to imagine "what a sound looks like" or "how a colour smells" (notwithstanding hallucinatory experiences of synestetic nature, e.g. a sound **stimulus** which evokes colour **experience.**

V. As to the **content** of conscious experience, one may very broadly speaking mention two classes of mental experience: mental objects which are related to sensory information from the external world and feelings, emotions or motivational states, which represent changes in the physiologico state of the organism.

Mental objects may be said to correspond to an internal "symbolic mapping" of the external world. They are not simple projections of corresponding external objects, but **sophisticated cognitive constructs**, based on selective information from the external world. One has to remind oneself that only a fraction of existing information is available for our senses. Also, there is an integration of **temporal sequences** of information into a single "momentary sensation". Perception is defined as the interpretation of an actual sensory message in terms of earlier experiences. More than 35 years ago the Cambridge psychologist, Kenneth Craik, suggested that our perception is a model about our world, presupposing the categories of space and time (Craik, 1967; Blakemore, 1977). In fact, the brain shows innate mechanisms which, to a large extent, determine the way we perceive our world. One could mention depth perception, constancy of perceived objects, a tendency to perceive patterns in terms of simple "gestalts", or a preference of a three dimensional interpretation of our world. The following two–dimensional figures are simple illustrations of this fact:

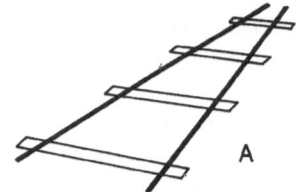 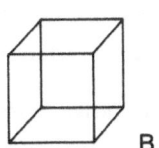 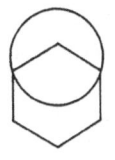 

A   B   C

Riedl (1979) gives an interesting discussion of these phenomena as to their biological significance.

Bower (1966) has shown the existence of elaborate visual mechanisms (binocular and motion parallax, constancy phenomena, gestalt perception etc.) at a very early age (2 – 8 weeks). Also, the brain seems to show innate expectations regarding the solidity of three-dimensional objects (Bower, 1971).

Nevertheless, the internal creation of cognitive constructs is based on a process of reconstruction where lost or otherwise missing sensory information may be regained in a hypothetical way. This demands the existence of earlier stored knowledge. Actual perceptions, experience and scientific models share a common feature in that they are related to knowledge in the sense of "insight" (Erkenntnis). The internal reconstruction of mental objects in connection with perception not only uses selected information, but the process is generally unconscious, uncritical and non-revisable, but the result is instead visualizeable. Now, the brain has the ability to transcend these limitations stepwise. One may think of the brain's ability to transform and handle mental constructs, e.g. the movement or rotation of a thought object in mental space. In this way knowledge by experience is created with the aid of conceptualization, language and logic. It means that consciously controlled, but still uncritical reconstructions are performed. Scientific knowledge, finally is based on highly abstract concepts and formalized models (black holes, neuron stars, particle-wave-dualism, quarks etc.); reconstruction is critical, but often no longer visualizeable (Vollmer, 1982, 1984). The above described functions of our brain may be said to constitute the background of what in evolutionary epistemology is called hypothetical realism.

Finally, it may be added, that the activities of the brain are highly integrated, which means that our subjective experience generally comprises both cognitive (symbolic) and emotional elements simultaneously. This implies that all brain structures, including subcortical structures such as the hypothalamus, limbic system, brain stem etc. contribute to the total experience. It does not only mean a superposition of mental constructs and moods, emotions and motivational stated, but also that the latter exert control functions on sensory filter mechanisms (direction of attention) which selectively influences the input of information. Attention however, is also controlled by acts of will, demanding the coordination of arousal and activation by exertion of effort (see Pribram & McGuiness, 1975).

VI. Self-consciousness did probably arise from the necessity of an organism to make a distinction between itself and the environment. A simple pre-stage is the ability to recognize that touching one's own organism (body) gives rise to signals from two different points, both of which represent the organism itself. Higher organisms then developed the ability to create an internal cognitive representation of themselves in toto. But there exist probably different levels of this ability. While animals, such as cats, may react to their mirror image as a signal which indicates another cat, human beings recognize the image as a symbol representing themselves. At least humans have an awareness of being aware. This seems to be the essence of Descartes`s "cogito, ergo sum". Humans perceive themselves as feeling and sensing entities of continuous existence during a limited period of time. It is the identification of "self–consciousness" with a sensing entity, denoted "ego" or "self" which makes it easy to slip into the fallacy of an internal homunculus ("the little spiritual man in the outer man"); (Crick, 1979).

VII. One may finally as,, why consciousness arose at all during phylogenetic evolution? There are several answers, which, as yet cannot be proven conclusively. One possibility is, that the emergence of consciousness is inevitable (due to its existence as a universal potentiality) as soon as the necessary preconditions are met. The brain is mainly an instrument for survival; its ability to receive and handle information is essentially a tool for this purpose. Conscious information processing represents a principle which combines high efficiency regarding sensory discrimination with economy regarding the necessary amount of neural elements. A hypothesis of this kind was expressed back in 1963 by Culbertson (p.79 and chapter 7). A principle which represents superior qualities is likely to be preserved in phylogenetic evolution and will probably be "adopted" by all organisms which can make use of it. The existence of such a principle would mean a strengthening of the organisms "negentropic flexibility" and this would have survival value for the organism. Sayre expresses this view in the following way: "Human consciousness, like learning is the product of evolutionary bia towards life forms with superior negentropic flexibility, for conscious organisms excel in their ability to receive information and to apply it under a wide variety of living conditions. Like learning, also consciousness is a form of adaptation parallel to the evolutionary process in its feedback characteristics" (Sayre, 1976), p. 139).

To be more specific: any organism which is confronted with the problem of coping with its adaptation to a very complex environment not only has to integrate a large amount of different information, but this integration also has to be done in a way which makes it possible to optimize a holistic response. Such a task includes, for example, the probability matching of different sensory messages in space and time in order to reveal their possible causal relationship.

The performance of an appropriate behavior within the given context has to be based on a processing of information which includes an internal representation of the organism itself together with elements of its environment. An animal that successfully wants to jump over a cliff, has to judge its own trajectory in relation to distance and initial acceleration. Hence both information about the environment and the organism itself has to be integrated into an internal "model" or program which determines complex actions. It is probable that "conscious" modes of information processing have developed in relation to this type of task which demands the simultaneous integration of large amounts of information into a single pattern. A detailed discussion of this topic from a phylogenetic point of view is given by Jerison (1978).

The last paragraph has briefly dealt with some factors which may have contributed to, or been responsible for, the development of conscious information processing.

While the topic is much more comprehensive, only a few further factors will be treated under the heading of the next paragraph.

8. MIND AND BRAIN.

The body-mind problem is well-known from the history of philosophy, where its classical treatment is part of ontology. The well-known clock analogy, which is normally used to illustrate Leibniz` principle of pre-established harmony, may serve this purpose for other monistic and dualistic ontologies as well. For reference see, for example Leahey (1980) or Vollmer (1980). The present paper does not deal with the problem of ontological reduction, which means that the general ontological notion of a "substance" can be excluded from the present context. If one still wants an analogy, the given model may be represented by a clock with one clockwork, but two faces with different presentations of the same information.

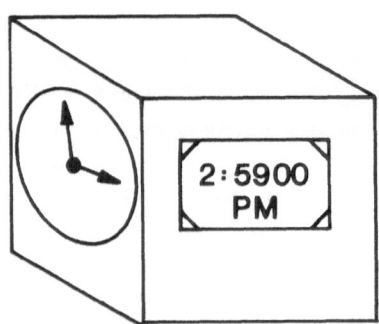

The problem then is "reduced" to the relationship between "brain and mind" as functional correlates, or more precisely, that neurophysiological events and their conscious attributes (mental events) are correlated, since both are aspects of the same brain functions. In psychology, the body-mind problem results in an attempt to validate empirically the hypothesis of psycho-physical parallelism (both in its monistic and dualistic variety). While G.T. Fechner (1801 - 1887) tried to give mathematical proof to the identity hypothesis of Spinoza, E.H. Weber (1795 - 1878) tried to quantify the rules of perceptual experience against the background of Leibniz theory of monads. The results are well-known as "psycho-physical laws". i.e. empirical generalizations of observed perceptual performances. An example is there Weber ratio:

$$\triangle R = \triangle S/S = k$$

where $\triangle R$ = "the unit of a subjective intensity scale" $\triangle S$ = the just noticeable difference of physical intensity (JND) against the initial value S. Fechner supposed a linear additivity of the units $\triangle R$. If $\triangle R$ and $\triangle S$ are viewed as approximations of differentials ($\triangle R$ and $\triangle S$),   a formal integration gives the Fechner Law:

$$R = k. \log S$$

S.S. Stevens (1906 - 1973) doubted the assumptions of a constant value $\triangle R$ and its linear additivity, which had not been proven. He proposed instead   a power law, based on direct comparative judgements:

$$R = k. (S - S_o)\eta$$

(S_o is a treshold correction).

To give an example: For a sound of frequency of 1000 Hz, the relationship between loudness (in sone units) and physical intensity (in dB) is:

$$R = k. S^{0.3}$$

The equation says that an increase of the sound intensity S by a factor 10 gives an increase of perceived loudness by a factor 2. An advantage of Steven`s Power Law, is that a logarithmic transformation gives a linear relationship, which makes comparisons of different situations easy.

The body-mind problem may, within the realm of psychophysiology, be viewed as a generalized psychophysical problem. This, however, soon leads to difficulties. The first problem is that classical psychophysics is not sufficiently applicable to everyday reality, since one must always consider the presence of random noise, both external noise and internal "dark noise" from the CNS itself. An attempt for the shortcomings of earlier models led to the development of Signal Detection Theory. Essentially it deals with the ability of an observer to detect signals against a background of random noise (Swets, 1964). The efficiency of an observer is described in terms of conditional probabilities, related to binary events (yes/no; signal/no signal).

A new difficulty is indicated if one tries to establish a relationship between the frequency of sounds and perceived pitch. To begin with, a difference between the eye and the ear must be considered. While the eye integrates complex stimuli (such as a mixture of light with different wave lengths) into a single colour perception, the ear analyzes a composite sound in terms of its components. A mixture of monochromatic "red" and "green" light of equal intensity is perceived as "yellow".*) The ear, however, performs a Fourier analysis of a complex sound with respect to its harmonic components. This phenomenon is called Ohm's Acoustic Law (Georg Simon Ohm, known from the theory of electricity). This property of the ear makes its possible for us to listen to polyphonic music or to hear the timbre of sound.

If the frequency of a sound is doubled. e.g. from 440 to 880 Hz, we perceive this jump as an octave. The resulting linear relationship between frequency and pitch is called the "musical scale". It is partly the result of learning (relating to the "well tempered clavichord" of J.S. Bach). Thoroughly controlled laboratory measurements. however. give a slightly different picture. Up to a frequency of 500 Hz there exists a linear relationship between pitch (in "mel" units) and frequency (in Hz). Above 500 Hz, the relationship is logarithmic:

$$z \text{ (mel)} = 500 + 1,300 \log f/500$$

If we relate perceived sounds to both intensity and frequency, the picture gets much more complex, since the sensitivity of the ear varies in a non-linear manner as a function of frequency.

The picture becomes still more complex if we regard colour perception. Colours are described in terms of hue, saturation and brightness. These dimensions are. however not independent, which is indicated by the Bezold-Brücke effect (cp. Day, 1969), which shows, that a monochromatic light subjectively changes its hue if the intensity (brightness) is changed. Exceptions are the three so-called primary colours of the Young-Helmholtz' Three Component Theory of Colour Vision. This

*) "Red", "green" and "yellow" are used here according to common language as labels to denote the dominant wave lengths. In a more strict language the wavelength numbers should be given. Also there often is a misunderstanding as to the actual meaning of the term "colour". The retinal receptors have different spectral sensitivity to light of different wave length. This is distinct from their ability to mediate colour information.

theory already clearly states that colour perception is a pure subjective phenomenon, which reflects properties of our perceptual mechanism.

The above mentioned factors render it impossible to arrange perceived colours within an Euclidean space. This is due to the fact, that the above mentioned dimensions of description are interdependent. The arrangement of colours in three-dimensional colour space can, however, be achieved in terms of non−Euclidean Riemann−geometry.*) Erwin Schrödinger (1920) has given a detailed treatment of the subject. A comprehensive survey, dealing with additional questions, has been given by Max Born (1963).

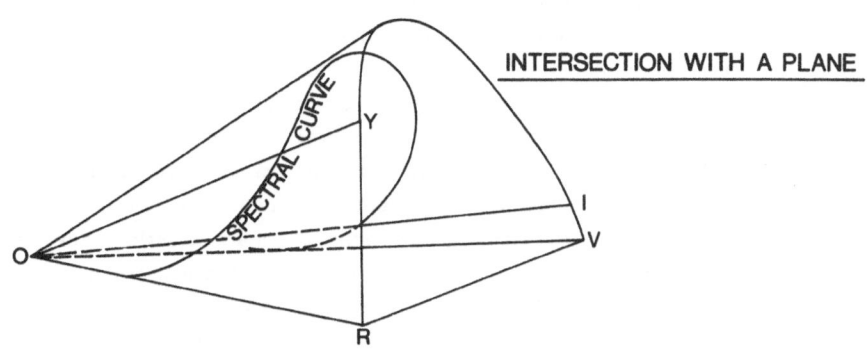

If one wants to apply the original psychophysical method to colour perception, it breaks down entirely. This is due to the fact that the brain transforms **quantitative** attributes of the stimulus into **qualitative** attributes of the corresponding percept. Monochromatic light is defined by its wave length (or photon energy), i.e. single, quantitative measures. But what are the **quantitative** measures of colours? Obviously. there are no such measures.

*) Spherical geometry, known from navigation and astronomy, is an example of Non−Euclidean geometry.

physical scale – subjective scale	physical aspect – subjective aspects
	(of coloured light)

τ_1	...........................Red	dominant wave	..................hue
τ_2	...........................Orange	length	
τ_3	...........................Yellow	spectral	.............................saturation
τ_a	...........................Violet	purity	
		light intensity	..................brightness

This fact is also reflected in language, which makes it necessary to describe the two aspects of light in a different terminology. The principle of colour experience illustrates nicely the efficiency of conscious information processing. The discriminative power of the principle is evident from the fact that the normal eye is able to discriminate about $7. \ 10^6$ different shades in direct comparison. This result is obtained by stepwise information processing. The eye contains three types of spectral sensitive receptor elements (cones). Their spectral sensitivity characteristics have maxima at about 445 nm, 535 nm and 570 nm respectively. The elements may be described as equivalents to band pass filters within the renege of visual light with partially overlapping characteristics. Edwin Land pointed out that colour perception demands a contrast situation, i.e. a comparison of signals from different receptors within the same receptive field (Crick, 1979). Hence, colour perception is based on competing processes within nerve nets. The nature of these processes, in turn, is determines by the relative intensity of signals from the three wavelength discriminating receptor elements (cones); (for reference see Biernson, 1968; Teevan & Birney, 1961).

The balanced interaction of these signal inflows (both excitatory and inhibitory) determines a very large number of different states of the nerve net and, hence, corresponding states of conscious experience. Meanwhile, how the latter comes about is part of the mystery of conscious experience.

The given examples show that basic conscious experience is due to **intrinsic** properties of the brain, while it is the role of receptors to establish a **transfer** function which gives the necessary coupling to external patterns of stimuli.

Two final questions should be mentioned. One is related to the autonomy of brain functions, and the other to the stability of our cognitive perceptual constructs.

The brain's ability to program itself is, like learning, an expression of its salient functional plasticity. The way in which this is achieved makes any distinction between software and hardware rather subtle. As to its autonomy, the brain runs without anyone telling it how to run. There are, of course genetically specified mechanisms of memory and perception (for reference see John, 1968; Pribram, 1969,

1971). One may mention the retinal organization in terms of receptive fields which relate to visual neurons at different hierarchical levels. The well-known works of Kuffler, Hubel and Wiesel as well as others revealed the existence in the retina of receptive fields which are sensitive to the orientation and movement of linear stimuli. Furthermore these studies showed the functional architecture of the area striata (visual cortex) in terms of cortical modules or columns. Each column contains many thousands of neurons which all are related to receptive fields of a specific orientation, different ones for different columns (Kuffler, 1953; Hubel, 1963, 1979; Hubel & Wiesel, 1960, 1962, 1968, 1979).

Lately, however, doubt has been expressed whether cortical neurons are solely "orientation detectors" for light or dark "bars" projected onto the retina. Rather they may be looked upon as spatial frequency filters performing a total Fourier analysis across a strip of the visual world where retinal fields of different orientation may be viewed as narrow windows towards the external world. A multichannel spatial filter model has therefore been proposed by DeValois (1980). Also Lassonde, Pribram & Ptito (1981) have shown that visual contical neurons cannot be classified with respect to single properties, but that their receptive fields represent transfer functions which express multiple feature selectivities.

Nevertheless, there is the problem of recognition and "meaning". In the simplest case, these tasks may be related to mechanisms which utilize triggers, filters and templates. One may these ask if there is a template-matching with respect to memorybound templates of visual patterns or if there exists a stepwise feature analysis? Probably both principles coexist; (for a discussion of the question see Lindsay & Norman, 1973). In connection with an application of Information Theory to human communication, Warren Weaver discusses the three levels of communication problem: the technical problem of accuracy with respect to the transmission of a message, the semantic problem of how well the transmitted symbols convey the desired meaning and finally the efficiency of the received message to affect the behavior of the receiver (Shannon & Weaver, 1959). A slightly different aspect is given by Hofstadter (1979), who speaks of three levels of the message: The understanding of the frame message means the recognize the need for a decoding mechanism. The understanding of the outer message is to recognize the code or to know, how to gain access to the correct "decoding mechanism" (translation of a language or symbols). This, finally, gives access to the inner massage i.e. the meaning of the message, intended by the sender.

The brain is confronted with similar problems in creating meaning (here in the sense of recognizable order). It is one of the problems of evolutionary epistemology to provide an answer to the question how the mechanisms which underlie this task have evolved. Evidently, these mechanisms have to be phylogenetically a posteriori, but ontogenetically a priori (in the sense of genetic potentials). The brain seems to be an open system with partially stochastic working principles (Scholl, 1956). Brain functions have also been described in terms of hierarchical random nets (Rapoport & Shimble, 1948; Rashevsky, 1960). Griffith (1971) additionally includes a field theoretical approach. In connection with certain types of tasks brain functions have been described in terms of associative networks with partly stochastic interaction. This means that the system chooses the most efficient coupling between a set of input- and output channels which fulfils the requirements of the actual problem (Crick, 1979). An illustrative example regarding the development of stereoscopic vision in kittens is given by Guillery (1974) and Stent (1975). Similar network theories with respect to overload and parasitic (unwanted) modes of excitation have been proposed as explanations for dream activity (Crick & Mitchison, 1983; Melnechuk, 1983).

The fact that the problem of internal order is related to the problem of classification was early recognized by von Hayek (1952). It should, however, be mentioned that affective classification as a means of concept formation in both animals and humans plays an essential role in Eugenio Rignano's psychology of reasoning (Rignano, 1923). Edelman (1982) now presents a neurophysiological selection theory of higher mental functions, which presupposes the existence of different neural groups, forming isofunctional, but non-isomorphic networks which constitute a selective system of degenerative groups. Similar acts of perception create a strengthening of connections between participating nets.

A final question regards the stability of the brain's performances. Despite its partially stochastic interconnections, the behavior is neither totally random nor chaotic, at least very seldom (epileptic seizures). The study of periodically forced, non-linear oscillators, showing period doubling and finally chaotic dynamics as a function of a few parameter values has during the recent years gained attention (see e.g. Perez & Glass, 1982). An application of similar models to nerve nets has been studied by several authors (see e.g. Holden et al. 1982; Labos. 1984).

From what has been said earlier, it is clear that the functional elements of the CNS are nerve nets, while neurons are structural elements. As to the basic stability of the brain, it is probably dependent on the fact that the parameters which govern the activity of each neuron have suitable values within a safety range, which normally prevent chaotic behavior.

As to observable phenomena. which reflect innate mechanisms that stabilize our perceptions, one may mention

*   sense modalities, which are non-interchangeable,

- the preference to see three-dimensional objects and an innate expectation of their solidity,

- gestalt perceptions and spontaneous organization of elements into simple patterns

- the multistability of perceptually unstable organizations (ambiguous figures). Examples are two-dimensional projections of transparent cubes (see earlier picture) or symmetric silhouettes. The essential point is, that the brain alternates between equally probable stable interpretations, but does not convey an unintelligible intermediate picture (Attneave, 1971).

- A final example is the mechanism behind perceptual constancy. While the shape and size of the retinal image may vary due to the position of an observer and the colour of an object may vary due to a variation of illumination, the size, shape and texture of the object are perceived as being constant. One supposes that shape and size constancy are dependent on mechanisms which essentially analyze the relative movement of pictorial elements, in the simplest case of dots. Elements which show a common motion are perceived as forming a group. A systematic movement of some elements relative to others are perceived as a motion of a subgroup within a common group. An example is the movement of legs and arms in relation to a walking person. Essentially the mechanisms seems to perform an analysis with respect to the degree of coupling between the velocity vectors in space for each element of the group. In terms of this analysis, which is very rapid, the consecutive retinal pictures may be said to form geometrical transformation groups. Size constancy then corresponds to the group of equiform transforms and shape constancy to the group of projective transforms (see Bergström & Jansson, 1977; Johansson, 1964, 1975).

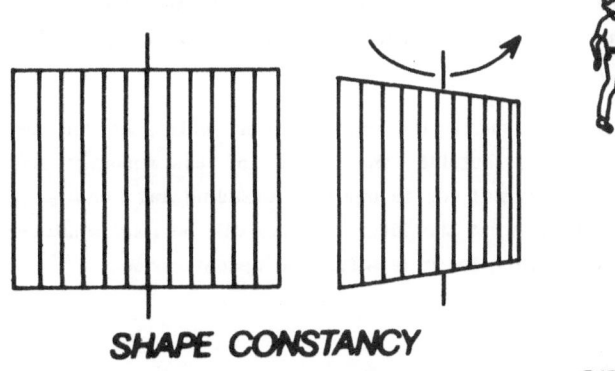

SHAPE CONSTANCY

SIZE CONSTANCY

The immensity of the topic leaves many questions open, which could not be dealt with explicitly. One may mention the essential role of memory in connection with conscious reconstruction, images and remembering. There is vast literature on the topic. From both general and neurobiological points of view the subject is treated by Bartlett (1961), John (1968), Lewis (1979), Pribram (1969, 1971), to mention a few authors. A further question of importance which relates to conscious brain functions is the problem of hemispheric specialization (lateralization). As known, both cerebral hemispheres contribute to an integrated mind by an exchange of highly specific neural codes which are mediated by the corpus callosum (the great cerebral commissure). For reference see Gazzaniga (1967); Gazzaniga & Le Doux (1978); Ornstein (1977); Sperry (1964).

The brain still holds many secrets which are difficult to understand due o a lack of detailed knowledge. Multiple personality is one of them. A person, during different periods of her life, may show quite different personalities (which are unaware of each others existence; Hilgard, 1977). This means the storage in memory of complete programs representing different personalities whose emergence, however, is state dependent. While both sets of programs are stored in the same substrate, they are only available under appropriate states. The fact that one "personality" is not even aware of the existence of the other, gives an extra dimension to the issue of consciousness.

As to the correlation between "mind" and "brain", it is quite clear from what has been said earlier that the problem is outside the range of classical psychophysics and simple correlations. One has to introduce multiple staged information processing models which allow for multiple interactions between factors. Linear, multivariate statistical methods (multiple correlations, eigenvalues and principal factors) are already commonplace in scientific psychology. The access to powerful computers has made them feasible. However, the more recent results from the study of chaotic systems, from synergetics, autocatalytic processes and non-linear feedback may make it necessary to introduce non-linear models as well. This will add to the technical difficulties.

The brain is a marvellous instrument. It shows features, which, as yet, are not understood and may still be outside the range of contemporary science. Nevertheless, any new discovery will add to our understanding and slightly change the picture. We can only guess, what the future picture will be like. Whatever the ultimate answer will be, the search for it is a fascinating enterprise.

Bibliography

Arnal, D., Gerin, P. Comparaison des potentiels évoqués moyens à l'aide d'un test T³ généralisé. **Electroenceph. clin. Neurophysiol.** 1969, **26**, 325–331

Attneave, F. Multistability in perception. **Scientific American**, 1971, **225** (6), 63–71.

Aurell, C.G. Perception: A model comprising two modes of consciousness. **Perceptual and Motor Skills**, 1979, **49**, 341–444.

Aurell, C.G. Perception: A model comprising two modes of consciousness. Addendum: Evidence based on event-related potentials and brain lesions. **Perceptual and Motor Skills**, 1983, **56**, 211–220

Ayala, F.J. The mechanisms of evolution. **Scientific American**, 1978, **239** (3), 38–61.

Barlett, F.C. **Remembering: A study in experimental and socialpsychology.** Cambridge: The University Press, 1961.

Bartley III, W.W.: The challenge of evolutionary epistemology. **Proceed. of the Eleventh Internat. Conf. on the Unity of Sciences.** Philadelphia, Nov. 1982, 835–880.

Bergström, S.S., Jansson, G. (eds) Contemporary research in visual space and motion perception. **Scandinavian Journal of Psychology**, 1977, **18** (2).

Bertalanffy, L. **Problems of life: An evaluation of modern biological thought.** New York: Wiley, 1951.

Bertalanffy, L. **General System Theory.** Middlesex: Penguin University Books, 1973.

Biernson, G.A. A review of models of vision. In: Levine, S.N. (ed). **Advances in biomedical engineering and medical physics, 2,** New York: Interscience Publ. (Wiley & sons), 1968.

Blakemore, C. **The mechanics of the mind.** London: Cambridge University Press, 1977.

Born, M. Betrachtungen zur Farbenlehre. **Jenaer Rundschau**, 1963, **8** (6), 235–248

Bower, T.G.R. The visual world of infants. **Scientific American**, 1966, **215** (4), 80–88.

Bower, T.G.R. The object in the world of the infant. **Scientific American**, 1971, **225** (4), 30–38

Brillouin, L. **Science and information theory.** New York: Academic Press, 1962

Bunge, M. The metaphysics, epistemology and methodology of levels. In: **Method, model and matter.** Dordrecht: Reidel, 1973 pp. 160-168.

Bunge, M. Is chemistry a branch of physics? **Zeitschrift für allgemeine Wissenschaftstheorie.** 1982, 13 (2), 109-223.

Butler, P.J., Klug, A. The assembly of a virus. **Scientific American,** 1978, 239 (5), 52-59

Campbell, D.T. Evolutionary epistemology. In: Schilpp, P. (ed). **The philosophy of Karl Popper** Vol. I. Lasalle: Open Court, 1974, pp. 413-463.

Carrel, A., Lindbergh, C.A.**The culture of organs.** London: Harper & Harper Broth., 1938.

Craik, K. **The nature of explanation.** London: Cambridge University Press, 1967.

Crick, F.H.C. Thinking about the brain. **Scientific American,** 1979, 241 (3), 181-188.

Crick, F.H.C., Mitchison, G. The function of dream sleep. **Nature.** 1983, 304, 111-114

Culbertson, J.T. **The mind of robots.** Urbana: University of Illinois Press, 1963

Culbertson, J.T. **Consciousness: Natural and Artificial.** The physiological basis and influence on behaviour of sensations, percepts, memory images and other mental images experienced by humans, animals and machines. Roslyn Highs, N.Y.: Libra, 1982.

De Valois, R.L.. De Valois, K.K. Spatial vision. **Annual Review of Psychology,** 1980, 31, 309-341.

Day, R.H. **Human perception.** London: Wiley & Sons, 1969.

Dickerson, R.E. Chemical evolution and the origin of life. **Scientific American,** 1978, 239 (3), 62-78.

Dobzhansky, T. **Evolution, genetics and man.** New York: Wiley, 1955.

Eccles, J.C. **Facing reality.** Basel/Heidelberg/Berlin: Springer (Edition Roche), 1970.

Edelman, G.M. Through a computer darkly: Group selection and higher brain functions. **Bull. Amer. Ac. of Arts and Sciences,** 1982, 36 (1), 20-49.

Eigen, M. Self-organization of matter and the evolution of biological macromolecules. **Naturwissenschaften**, 1971, 58, 465–523.

Eigen, M., Schuster, P. The hypercycle. A principle of natural self-organization. **Naturwissenschaften**, 1977, 64, 451–565. **Naturwissenschaften**, 1978, 65, 7–41

Eigen, M., Winkler, R. Das Spiel. Naturgesetze steuern den Zufall. München/Zürich: R. Piper, 1975.

Ehrensvärd, G. **Liv, ursprung och utformning.** Stockholm: Aldus, 1961.

Fermi, E. **Thermodynamics.** New York: Dover Publications, 1956.

Fiddes, J.C. The nucleotide sequence in a viral DNA (OX 174). **Scientific American,** 1977, 237 (6), 54–67.

Fulton, J.F.A. **A textbook of physiology.** Philadelphia/London: W.B. Sounders, 1956.

Gazzaniga, M.S. The split brain in man. **Scientific American**, 1967, 217 (2), 24–29

Gazzaniga, M.S., LeDoux, J.E. **The integrated mind.** New York/London: Plenum Press, 1978.

Gemme, G., Bernhard, C.G. **Ögats funktion hos djur och människa.** Stockholm: AWE/Gebers, 1975.

Glansdorff, P., Prigogine, I. **Thermodynamic theory of structure, stability and fluctuation.** New York: Interscience Publishers (Wiley), 1971.

Griffith, J.S. **Mathematical neurobiology.** London/New York: Academic Press, 1971.

Guillery, R.W. Visual pathways in albinos. **Scientific American**, 1974, 230 (5), 44–54.

Gunn, J.E. **The listeners.** New York: Charles Scribner's Sons, 1972.

Gurney, R.W. **Elementary quantum mechanics.** Lincoln: University of Nebraska Press (reprint; originally published by Cambridge University Press, 1935).

Haken, H. **Synergetics.** New York: Springer, 1976.

Halstead, L.M. **The evolution of the dinosaurs.** London: Peter Lowe, Eurobook Ltd, 1975.

Hayek, F.A. von: **The sensory order.** London: Routledge & Kegan Paul Ltd., 1952.

Hilgard, E. **Divided consciousness: Multiple controls in human thought and action.**
New York: Wiley, 1977

Hilgard, E. Consciousness in contemporary psychology. **Annual Review of Psychology,**
1980, **31**, 1-26

Hofstadter, D.R. **Gödel, Escher, Bach. An eternal golden braid.**   Harmodsworth,
England: Penguin Books, 1979.

Holden, A.V., Winlow, W. Haydon, P.G. The induction of periodic and chaotic
activity in a molluscan neurone. **Biol. Cybernetics,** 1982, **43**, 169-173.

Hubel, D.H. The visual cortex of the brain. **Scientific American,** 1963, **209** (5), 54-
62.

Hubel, D.H. The brain. **Scientific American,** 1979, **241** (3), 39-47.

Hubel, D.H. Wiesel, T.N. Receptive fields of single neurons in the cat's striate
cortex. **J. Physiology.,** 1960, **150**, 91-104.

Hubel, D.H., Wiesel, T.N. Receptive fields, binocular interaction and functional
architecture in the cat's visual cortex. **J. Physiology,** 1962, **160**, 106-154

Hubel, D.H., Wiesel, T.N. Receptive fields and functional architecture of monkey
striate cortex. **J. Physiology,** 1968, **195**, 215-243.

Hubel, D.H., Wiesel, T.N. Brain mechanisms of vision. **Scientific American,** 1979, **241**
(3), 150-162.

Jaspers, K. **Allgemeine Psychopathologie.** Berlin: Springer, 1948.

Jerison, H.J. The evolution of consciousness. **Proc. of the Seventh Intern. Conf. on
the Unity of the Sciences.** Boston, Nov. 1978, 711-723.

Johansson, G. Perception of motion and changing forms. **Scandinavian Journal of
Psychology,** 1964, **5**, 181-298.

Johansson, G. Visual motion perception. **Scientific American,** 1975, **232** (3), 76-88.

John, E.R. **Mechanisms of memory.** New York/London: Academic Press, 1968.

Kandel, E.R. Small systems of neurons. **Scientific American,** 1979, **241** (3), 66-76.

Kanitscheider, B. **Reduction and emergence in the unified theories of physics.**
Contr. to The Thirteenth Int. Conf. on the Unity of the Sciences. Washington D.C.
sept. 1984.

Kaplan, R.W. Der Ursprung des Lebens. Stuttgart: G. Thieme, 1978.

Krech, D,Crutchfield, R.S., Livson, N., Wilson Jr., W.A., Parducci, A. Elements of Psychology. New York: A. Knopf, 1982.

Kuffler, S.W. Discharge patterns and functional organization of the mammalian retina. Journal of Neurophysiology. 1963, 16, 37-68.

Labos, E. Periodic and non-periodic motions in different classes of formal neural networks and chaotic spike generators. In: Trappl, R. (ed). Cybernetics and Systems Research II. Amsterdam: North-Holland, 1984. pp. 237-243.

Lagerkvist, U. Gene Technology. Lecture given at The Institute of Physics, Chalmers University of Technology, Göteborg, Nov. 18, 1983.

Lassonde, M.C., Pribram, K.H., Ptito, M. Classification of receptive field properties in cat visual cortex. Experimental Brain Research, 1981, 43, 119-130.

Laszlo, E. Introduction to systems philosophy. New York: Harper & Row, 1973.

Leahey, T.H. A history of psychology. Englewood Cliffs: Prentice Hall, 1980.

Leinfellner, W. Evolution of intelligence. In: Weingartner, P. Czermak, J. (eds). Concepts and approaches in evolutionary epistemology. Vienna, 1983, 161-168.

Leinfellner, W. Evolutionary causality, theory of games, and evolution of intelligence. In: Wuketits, F.M. (ed.) Concepts and approaches in evolutionary epistemology. Boston: Reidel, 1984(a) pp. 233-276.

Leinfellner, W. Reductionism in biology and the social sciences. Contr. to the Thirteenth Internat. Conf. on the Unity of the Sciences, Washington DC, sept. 1984 (b)

Lewis, D.J. Psychobiology of active and inactive memory. Psychological Bulletin, 1979, 86 (15), 1054-1083.

Lindsay, P.H., Norman, D.A. Human information processing. New York/London: Academic Press, 1973.

Löwenhard, P. Consciousness - a biological view. Göteborg Psychological Reports, 1981, 11 (10).

Löwenhard, P. Knowledge, belief and human behaviour. Göteborg Psychological Reports, 1982 (a), 12 (11).

Löwenhard. P. Discussant's remarks (evolutionary epistemology). **Proc. of the Eleventh Internat. Conf. on the Unity of the Sciences.** Philadelphia, Nov. 1982 (b). pp. 901 – 906. A more extensive paper will be published during 1985.

Lorenz, K., Wuketits, F.M. (eds.) **Die Evolution des Denkens.** München/Zürich: Piper & Co. 1983.

Mayr, E. Evolution. **Scientific American,** 1978, **239** (3), 38–47.

McCulloch, W.S. Why the mind is in the head. In: Jeffres, L.A. (ed.) **Cerebral mechanisms in behaviour.** New York: Wiley, 1951, 42–57.

Melchenuk, T. The dream machine. **Psychology of Today,** 1983, Nov. 22–34.

Melzack, R. The perception of pain. **Scientific American,** 1961, **204** (2), 41–49.

Miller, J.G. **Living systems.** New York: Mc Graw Hill, 1978.

Mustelin, N. Livets utbredning i universum. **Kosmos,** 1983, **60,** 213–233. (Stockholm: Forskningsradens Förlagstjänst).

Nicolis, G. Prigogine, I. **Self-organization in non equilibrium systems.** New York/London: Wiley, 1977.

Nordling, C. Energi – en introduktion. **Kosmos,** 1982, **59,** 19–34. (Stockholm. Forskningsrådens Förlagstjänst).

Ornstein, R.E. **The psychology of consciousness.** New York: Harcourt, Brace Janovich Inc. 1977.

Perez, R. Glass, L. Bistability, period doubling bifurcation and chaos in periodically forced oscillators. **Physics Letters,** 1982, **90A** (9), 441–443.

Pierce, K. Putting Darwin back in the dock. "Scientific" creationists challenge the theory of evolution. **Time,** march 16, 1981, pp. 50–52.

Popper, K., Eccles. J. **The self and its brain.** Berlin: Springer, 1977.

Pribram, K.H. (ed.) **On the biology of learning.** New York/Chicago: Harcourt Brace & World, 1969.

Pribram, K.H. **Languages of the brain.** Englewood Cliffs: Prentice Hall, 1971.

Pribram, K.H., McGuiness, D. Arousal, activation and effort in the control of attention. **Psychological Review,** 1975, **82** (2), 116–149.

Prigogine, I. Order through fluctuation: self–organization and social systems. In: Jantsch, E., Waddington, H. (eds.): **Evolution and consciousness.** Reading, Mass.: Addison–Wesley, 1976.

Radner, D., Radner, M. **Science and unreason.** Belmont, Cal.: Wadsworth Publ. Co., 1982

Radnitzky, G. The science of man – biological, mental and cultural evolution. In: Cappelletti, V., Luiselli, B., Radnitzky, G., Urbani, E. (eds.): **Saggi Di Storia del Pensiero Scientifico dedicati a Valerio Tonini.** Rome: Societa Editoriale Jouvence, 1983.

Rapoport, A. Shimbel, A. Steady state in random nets I/II. **Bull. Math. Biophysics,** 1948, **10,** 220–226.

Rashevsky, N. **Mathematical biophysics.,** vol. II, ch. 20, p. 230–241.

Riedl, R. **Biologie der Erkenntnis.** Berlin/Hamburg: Parey, 1979.

Rignano, E. **The psychology of reasoning** (trans. W. Holl). London: Kegan Paul, Trench Trubner & Co., 1923.

Sagan, C. **The Dragons of Eden.** Ballantine Books, New York 1977.

Sanger, F., Air, G.M., Barrell, B.G., Brown, N.L., Coulson, A.R., Fiddes, J.C., Hutchinson, C.A. III, Slocombe, P.M., Smith, M. Nucleotide sequence of bacteriophage OX 174 DNA **Nature,** 1977, **265,** Feb. 24, 687–695.

Sayre, K. **Cybernetics and the philosophy of mind.** London: Routledge & Kegan Paul, 1976.

Scholl, D.A. **The organization of the cerebral cortex.** London: Methuen & Co., 1956.

Schopf, J.W. The evolution of the earliest Cells. **Scientific American,** 1978, **239** (3), 85–102.

Schopper, E. **Evolving physics and the problem of reduction.** (Discussion paper on Roman Sexl`s: Order and Chaos). Contribution to the Thirteenth Intern. Conf. on the Unity of the Sciences. Washington D.C., sept. 1984.

Schrödinger, E. Grundlinien einer Theorie der Farbenmetrik im Tagessehen. **Annalen der Physik,** 1920, **63,** 397–520.

Schrödinger, E. **Was ist Leben?** Bern: A Francke, 1951.

Sears, F.W. **Thermodynamics.** Reading, Mass.: Addison–Wesley, 1953.

Sexl, R.U. **Order and chaos.** Contribution to the Thirteenth Intern. Conf. on the Unity of the Sciences, Washington, D.C., sept. 1984.

Shannon, C.E., Weaver, W. **The mathematical theory of communication.** Urbana: University of Illionois Press, 1969.

Shimbel, A. Rapoport, A. A statistical approach to the theory of the nervous system. Bull. Math. Biophysics, 1948, 10, 41-55.

Sokolov, E.N. Introduction to learning in isolated neural structures (intracellular mechanisms of the associated learning) in: Adam, G., Mészáros, I., Bángai, E.I. (eds) **Adv. physiological science, 17, brain and behavior.** Budapest: Pergamon Press/Akademiai Kido, 1981.

Sokolov, E.N., Grechenko, T.N. Pacemaker plasticity in isolated neurons. In: Adam, G., Mészáros, I., Bángai, E.I. (eds). **Adv. physiological science, 17, brain and behaviour.** Budapest: Pergamon Press/Akademiai Kiado, 1981.

Sokolov, E.N., Willows, A.O.D., Concluding remarks on learning in isolated neuronal structures. **Adv. physiological science, 17 brain and behaviour.** Budapest: Pergamon Press/Akademiai Kiado, 1981.

Sperry, R.W. The great cerebral commisure. **Scientific American,** 1964, 270 (1), 42-52.

Stanley, W.H., Valens, E.G. **[Virus]** (trans. T. Warburton). Stockholm: Wahlström & Widstrand, 1964.

Stent, G. Explicit and implicit genetic information. **Proc. of the Fourth Internat. Conf. on the Unity of the Sciences,** New York, Nov. 1975. pp.261-277.

Swets, J.A. (ed.) Signal detection by human observers. New York: Wiley, 1964.

Teevan, R.C., Birney, R.C. (eds.) **Color vision.** New York: van Nostrand Co, 1961.

Uexküll, J. von **Theoretische Biologie.** Berlin: Springer, 1928.

Vidal, G. The oldest eukaryotic cells. **Scientific American,** 1984, 250 (2), 32.-41.

Vollmer, G. Evolutionäre Erkenntnistheorie und Leib-Seele-Problem. In: Böhme, W. (ed.) **Wie entsteht der Geist? Herrenhalter Texte,** 1980, 23, 11-40.

Vollmer, G. **Evolutionäre Erkenntnistheorie.** Stuttgart: Hirzel, 1981.

Vollmer, G. On supposed circularities in an empirically oriented epistemology. **Proc. of the Eleventh Intern. Conf. on the Unity of the Sciences.** Philadelphia, Nov. 1982 (a). pp. 783–833.

Vollmer, G. **Das alte Gehirn und die neuen Probleme. Aspekte und Folgerungen einer evolutionären Erkenntnistheorie.** Darwin–Symposium "Das Phänomen der Evolution". Vienna: Technische Universität, May 1982 (b).

Vollmer, G. **The unity of science in an evolutionary perspective.** Comm. to the Twelfth Intern. Conf. on the Unity of the Sciences. Chicago, Nov. 1983.

Vollmer, G. Reduction and evolution, arguments and examples. In: Balzer, W. Pearce, D., Schmidt, H.J. (eds.) **Reduction in science. Structure, examples, philosophical problems.** Dordrecht: Reidel, 1984 (a).

Vollmer, G. New problems for an old brain. Synergetics, cognition and evolutionary epistemology. In: Frehland, E. (ed.) **Synergetics – from microscopic to mecroscopic order.** Berlin: Springer, 1984 (b). pp. 250–258).

Weiss, P.A. The living system: determinism stratified. **Studium Generale,** 1969, 22, 361–400.

Weiss, P.A. **Hierarchically organized systems in theory and practice.** New York: Hafner Publ. Co., 1971.

Wiener, N. **Cybernetics or control and communication in the animal and the machine.** New York: MIT Press and John Wiley Inc. 1961 (2. ed.)

Wuketits, F.M. Das Phänomen der Zweckmässigkeit im Bereich der lebenden Systeme. **Biologie in unserer Zeit,** 1982a, 12 (5), 139–144.

Wuketits, F.M. Evolutionary epistemology, objective knowledge, and rationality: the evolutionary approach in man's search for himself. **Proc. of the Eleventh Int. Conf. on the Unity of the Sciences.** Philadelphia, Nov. 1982b, pp. 881–899.

Zethraeus, S. **Känslan.** Stockholm: Natur & Kultur, 1962.

Zurer, P.S. The chemistry of vision. **Chemical & Engineering News,** 1983, 61, 48, 24–35.

H. Mohr

IS THE PROGRAM OF MOLECULAR BIOLOGY REDUCTIONISTIC?

1. `REDUCTIONISM` AND `REDUCTION`: EXPLICATION OF TERMS TO BE USED IN THE PRESENT ARTICLE

(a) `Reductionism`, an integral part of the analytic-synthetic procedure of the natural sciences, is the practice of explaining the properties of wholes (a cell, an organism, a society, or some other complex entity) entirely by the properties of the parts that compose them. This obviously implies `analysis`. i.e. the breaking down of complicated systems into manageable components. However, analysis is only half of the reductionistic approach; the other half is `synthesis`, in which the relation of the parts, i.e. their spatial and temporal ordering and the way they interact, is also laid bare and the whole system then reassembled as a physical union (e.g., synthesis of a complex molecule in organic chemistry) or recreated in the mind of the scientist (e.g., a theoretical simulation of the whole with mathematical means). In biology, at least, analysis makes only sense if followed by synthesis; an in vitro study contributes to biology only if it fosters our understanding of the cell (or of higher levels of the hierarchy of animate nature, see Table 1).

Table 1. The various levels of the hierarchy into which the system of animate nature falls. A significant property of this arrangement is that the higher levels include the lower levels. However, empirical richness increases as we ascend the levels [8].

P. Hoyningen-Huene and F. M. Wuketits (eds.), Reductionism and Systems Theory in the Life Sciences, 137–159.
© *1989 by Kluwer Academic Publishers.*

Society/community
Organisms
Organs
Cells [a]
Cellular organelles
Macromolecules
Molecules
Atoms
Subatomic particles

[a] The cell is at the bottom of animate systems. Subcellular systems, e.g., a plastid, a gene or a biomembrane, cannot be considered alive. It is the cell which is alive, not their constituents. Thus, `life` as an emergent property can be recognized first at the level of the cell.

The reductionistic approach implies that all complex systems consist of smaller and simpler parts. Moreover, it is assumed as a matter of course, that complex systems originated from simpler systems in the course of a universal evolution. Evolution is considered a deterministic process, governed by causal laws, at least above the quantum level [1].

(b) The concept of `reduction` is simple in essence. It emphasizes the (metascientific?) question whether (and to what extent) a proposition, a theory, or a whole branch of science may be reduced to another proposition, theory or discipline [1]. However, as Vollmer pointed out recently, "the concept of `reduction` – originally hoped to be simple and clear-cut – turned out to be complicated and ambiguous" [1].

Among logical empiricists `reduction` as a normative postulate was mainly concerned with theories [2]. A theory T_2 can or cannot be reduced to a theory T_1. A reduction of T_2 to T_1 presupposes at least that (1) the terms of T_2 can be redefined in terms of T_1 and (2) that all general propositions (`laws`) of T_2 can be deduced from general propositions of T_1. Thus reduction is concerned with (theoretical) terms and laws (and thus implies a uniform scientific language, a problem by itself).

Reduction is not concerned primarily with the observational level. On the observational level biology and physics will always remain different sciences.

Does reduction actually work or is it a normative fiction of the philosophers of science?

In practice, the **tendency** for reduction has had a successful record in the history of the sciences. Reduction satisfies one of the great desires of the scientist – to have unifying theories with a wide scope. A typical example is the reduction of classical thermodynamics to statistical mechanics or the creation of a few theories of great generality such as quantum mechanics or the theory of relativity. These are impressive examples for successful reduction within the realm of physics. However, even here serious difficulties arise. Most scientists are convinced indeed that atoms and molecules are ruled exclusively by the known principles of electrodynamics and quantum mechanics as are Maxwell's equations, Schrödinger's (or Dirac's) equation, and Pauli's exclusion principle [1]. However, most atoms and all the more all molecules are too complicated for "ab initio" computations from principles: "Physicists have so far not been able to deduce the behaviour of an uranium or even an .oxygen atom" from the above principles, even though our knowledge as to these basic principles seems to be complete [1]. It is a pragmatic postulate to grant chemistry at least the status of a semi–autonomous field within physics. This implies a direct measurement of molecular properties (observational level) as well as the use of these data for theoretical statements that are not directly related or immediately translatable into the language of physics.

Within physics proper this sober pragmatism has been an essential feature in the difficult relationships between electromagnetism and mechanics. Faraday introduced the term (and the concept) of the electromagnetic field, and Maxwell developed the pertinent theory. As Heisenberg [3] pointed out, "... the physicists began to understand that a field of force in space and time could be just as real as a position or a velocity of a mass, and that there was no point in considering it as a property of some unseen material called either." However, the traditional belief among physicists that reduction to the mechanical concepts could finally be effected was a strong hindrance. As Heisenberg continues, "it was not before the discovery of relativity that the idea of the other was really given up, and thereby the hope of reducing electromagnetism to mechanics."

Is reduction desirable?

Irrespective of practical obstacles some philosophers [1] insist that "an ideal final stage of factual science would be a minimal, redundancy–free description of the real world." The regulative idea of such an enterprise will not only be truth, but also simplicity, economy, parsimony and elegance of propositions and theories. If we accept this position – for the time being – the question arises: Are there any arguments in favour of **feasibility** of a unified science in view of diverging disciplines and the excessive specialization in present–day science?

The major compelling argument stems from the concept of a universal evolution [1]: If complex systems originated from simpler systems according to causal laws, then it should be possible to explain the complex from the simple. This "argument from evolution" suggests that strict reduction between different disciplines and sciences should be possible, at least in principle. However, as Vollmer [1] points out himself, this `argument` may serve as a heuristic cue only pointing to the right strategy. It might well be that in practice the problems posed by the complexity of real systems and by the principal limitation of the human mind [4] will set narrow limits to reduction.

2. MEANING OF `MOLECULAR BIOLOGY`IN THE PRESENT CONTEXT

The term `molecular biology' designates a novel approach – both intellectual and technical – to explain heredity in molecular terms; not only formal transmission genetics but also developmental genetics, i.e. the translation of genetic information into traits and phenotype, the "mapping of genetic space into organismic space" [5].

The study of the machinery of protein synthesis and its regulation has been the core of `classical` molecular biology; the discovery of the DNA double helix in 1953, the notion of messenger RNA in 1960 and the deciphering of the genetic code in 1961 were early highlights in the history of this novel discipline.

Even though the experimental basis of `classical` molecular biology was dangerously small – restricted to relatively simple systems such as viruses, bacteriophages and bacteria (**Escherichia coli**) supplemented by transducing phages and sex factors –, it was felt that the elucidation of the mechanism of heredity and adaptive enzyme synthesis in procaryots was of general significance: "What is true for E. coli, is also true for elephants" (i.e. true for man), as Jacques Monod expressed the general conviction in 1962. The vectorial sequence DNA – RNA – protein was immediately and widely accepted as a `central dogma` – at least until it was shaken by the discovery of reverse transcriptases.

The stupendous success in the molecular analysis (and intellectual reassembly) of the information-processing machinery of the (bacterial) cell gave the molecular biologists the confidence to attack any biological problem. The protagonists clipped away at the fundamental tenets of biology aiming at a truly unifying theory of biology. No biological problem, including cancer, ontogenetic development in higher eucaryots, neurobiology, sociobiology, was considered too complex to resist the molecular biologist`s approach.

In the following the original optimism was further supported by fundamental new discoveries such as gene splicing, gene cloning and gene sequencing which led to practical methods for study of eucaryotic genomes.

With respect to the question of `reduction` the program of molecular biology has been strictly reductionistic from the very beginning until the present day. The goal has always been the unity of biological sciences: to explain the major elements of the classical biological disciplines in terms belonging to the most fundamental biological science, molecular biology.

The reductionistic scope was even wider: at least implicitly, molecular biology was considered by some leading heads a branch of physics [6,7]. For good reason: The notion of a universal evolution – cosmic evolution once started with a Big Bang and led eventually to organismic evolution, to conscious beings and to social systems [1] – implies that the physical and chemical properties of matter were sufficient to produce living systems. If this idea is accepted (there is no alternative available which would be compatible with factual science), we cannot but conclude that physical and chemical laws should be sufficient to explain the historical emergence and present existence of living matter, including conscious beings and societies. Clearly, for **scientific** reasons there is no **a priori** objection against `reduction`.

3. PRACTICAL LIMITS TO REDUCTIONISM: THE CONCEPT OF `EMERGENCES`

The traditional (superficial) question of the holist: Is the whole "a mere sum of its constituent parts?" is quickly answered: As a rule **not**, of course. In fact, very few properties of living systems can be represented as merely **additive** functions of the properties of constituent parts [8]. It is the functional relationship, the **interaction**, between the parts which matters. What we call "complexity" in biology, is due to the **peculiar** interaction of the parts (of molecules inside the cell, of cells within the organism, of organisms within a society). Regarding reductionism, most biologists` work is based on the conviction that the properties and performances of living organisms are and can be nothing but the performances of living cells gathered together into the state characteristic of living organisms [8]. The principle of reducibility is usually taken even one or two steps lower in the hierarchy (Table 1) – for cells can be resolved into molecules, molecules into their constituent atoms, and they in turn into elementary particles [8]. Since the natural sciences of today offer no support in favour of vitalism the hierarchy of complexity we observe in nature is no reason to argue against the concept of full-scale reduction.

Vitalism implies that some nonmaterial elements of a spiritual nature (called "force of life", "élan vital", or – with reference to Aristotle – "entelechies") are constitutive for living systems but not for nonliving systems. If vitalism means that a satisfactory explanation of living systems is beyond the capacity of factual science, then vitalism is fully obsolete. We do not know of any nonmaterial entities, restricted to living systems, that would principally forbid the reduction of biology to physics (down the hierarchy in Table 1).

However, it is evident that in practice reducibility has its limits. There are properties, called **emergent** properties, that appear at higher levels of integration and are not (yet) predictable or interpretable in the lower ranks of Table 1. The 'concept of emergence' its merely descriptive. It has no explanatory value but simply states our present ignorance. As an example, the statement that the possession of a mind in conscious beings is an emergent property, is descriptively correct: there is no psyche or rudiment of psyche detectable in a crystal or in a DNA molecule or in an elementary particle. But the apparently emergent property might well be considered a necessary property of any natural system once it reaches a high level of physical complexity [9]. Most molecular biologist probably share Medawar's view [8]: "The idea of emergence plays a useful part in the biological sciences if only by giving a name to that which does not (yet) respond to reductive analysis, though it must be added that biological research of all kinds prospers proportionately as more and more of its subject matter does respond."

Irrespective of theoretical playing down of emergences, however, in practical research `emergent properties` means limits to reductionsm! Nobody at the work bench expects seriously that study of a suspension of mitochondria – energy producing subcellular organelles – will yield the laws of energetic homoeostasis of the cell. We do not expect either, that even the most sophisticated molecular studies of plant cell suspension cultures will tell us the laws which govern developmental homoeostasis in plant embryogenesis. When working with fibroblasts we will not expect to discover those coexistence laws on which the mechanical stability of the human body is based, etc. In brief: emergences in practice unevitably lead to limits of reduction. To ignore the emergent properties at the different levels of complexity for the sake of maximum reducibility would mean to ignore the empirical richness of the animate world.

This tendency of ignoring the empirical richness of nature for the sake of efficient and far-reaching reduction has a long history in science. As an example, when I was introduced to biochemistry, one of my teachers – a Nobel laureate – told us "that in essence a cell is a little bag, filled with enzymes and substrates". This model of the cell allows indeed a far-reaching reduction of cell biology to chemistry but we all realize nowadays that it ignores most properties of 'real' cells.

We all believe that real cells exist. However, in biological theories the 'cell' is a construct, i.e. an intellectual invention that is useful in organizing the real world, and there are different models approximating this construct. These models look very different, depending on the interest of the investigator or the teacher. A cell model illustrating the coexistence law of water potential differs in appearance from the cell models used to emphasize the concept of apparent free space or the concept of membrane flow. All of them claim to represent the construct 'cell', which we believe to recognize when we look through the microscope [10].

4. TO WHAT EXTENT CAN MENDELIAN (CLASSICAL) GENETICS BE REDUCED TO PHYSICS?

It is a rule of formal logic, called the transitive relationship, that if A can be analytically reduced to B and B to C, then A can be reduced to C. This being so, the above question: To what extent can Mendelian genetics be reduced to physics, can be divided in three sub-questions (Table 2).

Table 2. **Upper part:** Some examples for successful reduction within physics.

Lower part: Three steps in process of reduction of classical (Mendelian) genetics to quantum physics (and electrodynamics) [10].

T_2: Theory to be reduced

T_2: Theory to be reduced	Classical thermodynamics	Newtonian mechanics
T_1: Reducing theory	Statistical mechanics	Relativistic mechanics
T_2: Classical genetics	Molecular genetics	Macromolecular chemistry
T_1: Molecular genetics	Macromolecular chemistry	Quantum theory
	Classical genetics	
	Quantum physics	

1.   Can Mendelian (classical) genetics be reduced to molecular genetics?

2.   Can molecular genetics be reduced to macromolecular chemistry?

3.   Can macromolecular chemistry be reduced to physics (i.e., to quantum theory and electrodynamics)?

I think that strict reduction (in Nagel's sense [2]) is not feasible in any of the three cases, at least at present.

ad 1. While nobody doubts that any Mendelian gene has a molecular equivalent (nominal definition: a structural gene is a segment of a DNA macromolecule that can be transcribed and that codes for a poly-peptide), the full-scale reduction of Mendelian genetics to molecular genetics would be a hopeless undertaking on the level of laws or law-like propositions at present. An example will indicate the crucial point: it is stated in nearly every text that Mendel's laws depend on the "random" distribution of the chromosomes during meiosis. If a diploid mother cell possesses 8 x 2 chromosomes, each of the four meiospores or gametes will have 4 chromosomes. This so-called "random" distribution of the chromosomes is, of course, not a statistical (or probabilistic) phenomenon but subject to a strict law, which can be formulated as follows: if one of the homologous chromosomes moves in one direction (toward one pole), the other one must move in the opposite direction (if no "error" or disturbance is involved). It is the `operation` of this law that guarantees an equal distribution of the chromosomes over meiospores or gametes. The only random event involved in meiosis is the decision as to which one of the two homologous partner chromosomes moves first toward a pole. If this decision has occurred, the partner no longer has any choice. In this way an apparently random distribution of chromosomes and genes can occur during meiosis even though the actual number of chromosomes involved, say $2n = 8$, would be far too small to achieve a random distribution by purely statistical means. In other words: a strict biological law, an emergent law, at the level of the eucaryotic cell, is the basis of the "random" distribution of genes during meiosis. This empirical law is also obviously an antecedent condition for population genetics, since it is the essential prerequisite for random behavior in recombination. There is not the slightest chance at present to reduce this law to any "law" of molecular genetics.

Another argument against reduction of classical genetics to molecular biology is that classical genetics is the only part of biology that has been successfully axiomatized by Woodger [11] using the formal apparatus of Whitehead and Russell's **Principia Mathematica.**

A similarly strict axiomatization has not been performed so far in the case of molecular genetics. With regard to axiomatization classical genetics are clearly superior to molecular genetics.

On the other hand, the scope of classical genetics has been much more narrow than the range of molecular biology. Classical genetics has been primarily transmission genetics; molecular biology intends to cover transmission genetics, including identical replication of genes, as well as developmental genetics. Developmental genetics is concerned with the causation of traits by genes during development. In this sense, classical genetics could indeed be regarded as a special, very limited case of the more general new theory. Any adequate gene model in molecular biology must satisfy the requirements of transmission as well as of developmental genetics. The Watson–Crick model of genetic DNA from 1953 [12] serves both functions. It not only explains the replication of genes but it is also consistent with the various models of gene function or gene expression, e.g., with the Jacob–Monod model from 1961 [13], which explains adaptive enzyme formation in bacteria. In the higher eukaryotes, which possess chromosomes and effector molecules such as hormones and phytochrome, the models to explain differential gene expression are more complicated but nevertheless consistent with the Watson–Crick model. All models for gene expression are based on the one-gene–one–enzyme hypothesis (today, one–cistron–one–polypeptide hypothesis) introduced originally by Beadle and Tatum in the forties and on the universal validity of the so–called genetic code, a set of rules of correspondence that determine the correct translation of a peculiar linear nucleotide sequence into a peculiar linear amino acid sequence.

These few statements may suffice to indicate the tremendous progress molecular biology has brought to biology. With regard to the problem of `reduction`, I feel that molecular biology will replace in practice classical transmission genetics as far as possible without any enforcement. Classical genetics will not be fully displaced since for many statements in transmission genetics the terms and laws of classical genetics are more useful than the corresponding statements of molecular biology.

ad 2: Can molecular genetics be reduced to macro–molecular chemistry? This is obviously the crucial question. Since molecular genetics (which always includes developmental genetics), presupposes the function of the cell, the question can be reformulated in more general terms: can the function of a cell or an organism be reduced to chemistry? There is no quick answer. The reason is that every living system is characterized by "organized complexity" whereby `organized` denotes that the components (elements, parts) of the system do not form a homogeneous, randomized aggregate. Rather, any living system is a highly improbable system that contains a large amount of "organizational information" (or, neg–entropy, spatial information, system information). Much of this specific, peculiar system information will be irreversibly lost if the system is destroyed. For example, if a cell is homogenized much of the system information is gone (some will be retained in the more or less intact organelles such as mitochondria or plastids). "They all look the same in the Waring blender" is an old saying of the (innocent) reductionists among the biochemists. This is precisely true: a yeast cell, a frog cell, or a human cell all look very much alike after homogenization. The crucial point, however, is that this biochemical advantage (e.g., for the study of basic cell components and fundamental cell functions) is paid for by the loss of essential information about the organization of the living system. In brief: analysis of isolated elements inevitably results in the loss of information about the particular and highly non-random relationships between the elements (parts) of the system.

Originally the organizational information was derived from genetic information. The kind of "self–assembly" from the parts, which works with the tobacco mosaic virus and partly with the T4–bacteriophage no longer works, for reasons of principle, on the cellular level. The construction of a protein molecule is a simple example of "directed assembly": the amino acid sequence is absolutely nonrandom. It is dictated by the nucleotide sequence of genetic DNA via messenger RNA.

The question (repeated once in a while) whether or not "life" can be created de novo in the test tube out of a soup of molecules is out of date. We know that the orderly construction of a living system depends not only on the availability of the physical elements (building blocks) and of free energy (power) but also on the availability of the pertinent information (blueprint). Moreover, the construction depends on those factors that can read the information and are able to guarantee an orderly use of the information (analogous to a fantastic master builder). At least in the foreseeable future the master builder must be obtained from pre-existing living systems.

In summary, a living thing cannot be explained in terms of its parts but only in terms of the organization of those parts. Although the whole is nothing but the parts put together, the "putting together" creates something that cannot be accounted for by the properties of the parts themselves. Every process occurring within living systems is (in principle) compatible with a physical mechanism. However, if we want to understand how the particular living system came into being we must refer to development guided by genetic information. It is obvious that only a very limited subclass of the possible interactions among cell constituents permits their integration into a living cell.

The analogy with teleological action, indicated in the foregoing section by the terms in parentheses, is obvious. A car is not more than the sum of the parts if (and this is the decisive point) the sum of the parts includes the spatial relationship (the "organization") of the parts. This organization has nothing to do with the parts themselves. It cannot be derived from the properties of the parts. It is specific and peculiar information that was introduced during the construction of the car. It can be described once the car is ready for use, as the system or organizational information, characteristic for this particular model of a car. Therefore one may not assume that the specific structure of a machine is explainable by the laws of physics [14]. The specific design is completely extrinsic. But, of course, every process occurring within the machine is compatible with a physical mechanism. In an analogous sense it is the specific and peculiar genetic information with its `historical dimension` that determines the specificity and the `design` of organismic organization. Only if evolution and ontogenetic development could be explained fully in terms and laws of physics, could an explanation sensu stricto of any real organism in terms and laws of physics be envisaged. This is the crucial point.

As stated above the notion of a universal evolution – a firm credo of science – implies that the physical and chemical properties of matter were sufficient to bring about living systems. This credo implies further [1] that if the physical and chemical properties of matter were sufficient to produce `life`, then physical and chemical laws should also be sufficient to explain the emergence of `life`[6,7]. At present we are indeed far from reaching this goal. However, it must be emphasized, that it is the practical (epistemological) limitation of the human mind that prevents full-scale reduction; there is no reason to suppose a principal, ontological barrier.

ad 3: The exact treatment of complicated aperiodic macromolecules with quantum mechanics still causes great difficulties. While these difficulties are not insurmountable in principle, in practice the theoretical treatment still depends on approximating methods that are not completely satisfactory and difficult to handle. We have considered this aspect in a previous section about the relationship between physics sensu stricto and chemistry.   Regarding molecular biology in particular, the interaction of large molecules with the watery media in which they are usually found is not so much a puzzle as an issue tacitly recognized to be too complicated to tackle and best for the time being ignored [15].

5. TO WHAT EXTENT IS `COMPUTER SIMULATION`A PROMISING MEANS
FOR REDUCTION IN (MOLECULAR) BIOLOGY?

A new kind of reduction, which has attracted much interest recently, is to simulate in a computer program some properties of a particular biological system. Simulation is a general term applied to the process of conducting experiments on a model instead of attempting the experiments with the real system [16]. In biology, simulation studies were made in particular with the use of computer simulation techniques. Computer simulation models are in a way intermediate between concrete empirical studies and the mathematical analysis of hypothetical, abstract models. In biology, simulation means setting up in a computer conditions that describe with reasonable approximation relevant properties of the biological system we want to study. On the basis of the input data (objective data as well as assumptions) the computer then generates the resulting information (output). Different input data and assumptions can thus be tested to determine their effects on the behavior of the system. As an example, the simulation of the process of glycolysis has already advanced quite far [17]. While the predictive value of simulation can be high at times, it is generally felt that simulation should only be regarded as an additional tool. It cannot be a substitute for an explanation in terms of the actual constituents of the living system.

With regard to psychology, the use of mathematical models and computer simulation could lead to a break-through [18]. It is generally accepted that neurophysiology is the material basis for psychology. However, a serious effort to reduce psychology to neurophysiology has not been made so far [19]. This kind of enterprise would first of all require a redefinition of many psychological terms in terms of neurophysiology. Many psychologists are obviously reluctant to consider seriously this formidable task, in particular since the entry of psychoanalysis into the academic scene. At this point, computer simulations of some functions of the human brain might be helpful. Most natural scientists feel that the human mind is "nothing more" than a correlate of a very highly organized material system, the human brain. In Crick`s (admittedly blunt) words: "I myself, like many scientists, believe that the soul is imaginary and that what we call our minds is simply a way of talking about the functions of our brain" [20]. This assumption implies that a computer of high system complexity approaching the material system properties of the human brain would show at least some of those properties that we usually attribute to the construct "mind".

6. TO WHAT EXTENT CAN `DEVELOPMENTAL BIOLOGY` BE REDUCED TO MOLECULAR BIOLOGY?

The declared goal of modern developmental biology is to describe the `mechanism` of development (ontogeny), i.e., the sequence of elementary steps and their regulation, in terms of molecular biology [21].

This implies an understanding of how the information encoded in the genes relates to the means by which cells assemble themselves into an organism in space and time. In development of multicellular organisms we observe not only the emergence of a diversity of cell types, but these differentiated cells are precisely localized within the developing embryo or organ. The resulting spatial organization can be referred to as the appearance of patterns: cells become arranged in regular, reproducible, non-random fashion to form specific tissues and organs. The mechanisms involved in the generation of patterns have intrigued investigators ever since developmental biology emerged. However, how these spatial and temporal patterns are formed is still one of the great enigmas of biology [22]. "It is a problem which has not, so far, benefited significantly from recent advances in the methodology of molecular biology. Instead, our efforts at understanding the formation of patterns in development are still comparable to the pre-Mendelian stage of genetics. We are still searching for the formal `rules` by which we can predict the behavior of embryos under various experimental treatments. Only when we understand these principles at a formal level, do we expect to be able to frame appropriate questions for a molecular analysis" [23].

A couple of years ago, the outstanding successes in analyzing the genetic material and the genetic code, and in uncovering the mechanism and control of gene expression in microbes tempted some leading molecular biologist (e.g. Crick and Brenner) to approach the new territory of developmental biology with the powerful tool of their scientific discipline. "The genetics and biochemistry of control mechanism in cellular development", was the chosen goal. It has proven to be a far less neat and tidy territory than the organized mind of the molecular biologist had contemplated [5]. "At the beginning it was said that the answer to the understanding of development was going to come from a knowledge of the molecular mechanisms of gene control", reflects Brenner today. "I doubt whether anyone believes that any more. The molecular mechanisms look boringly simple, and they don`t tell us what we want to know. We have to try to discover the principles of organization, how lots of things are put together in the same place. I don`t think these principles will be embodied in a simple chemical device, as it is for the genetic code" [5]. The reductionistic mindset has not been abandoned, of course; it is the impatience of the pioneer days of molecular biology which has been replaced by an appreciation of the complexity of the higher eucaryotic organisms. "Ultimately the organism must be explicable in terms of its genes, simply because evolution has come about through alteration in DNA. But the representation will not be explicit. We need to understand the 'grammar of development' to make sense of it" [5]. In particular we have to understand the principles of molecular assembly, the way the geometry of a structure is specified by the way the constituents interact during development. One must seek the grammar of assembly [5], the 'code of morphogenesis', even if the ultimate source of information is at the level of the gene. It is the **interaction** of the component parts of the system which determines 'morphogenesis'.

Plastidogenesis in higher plants is chosen to make the point.

Plastids are subcellular organelles in photosynthetic cells whose mode of appearance in the cell can be considered – from the point of view of the reductionist – as a model for 'morphogenesis' [24]. This implies that the basic questions related to morphogenesis of the organism (seedling) can in fact be studied at the level of a subcellular organelle.

In photosynthetic green plants light is the decisive environmental factor. The terrestrial green plant is organized almost ideally in a way so as to absorb and process light quanta. The genetic adaptation to the light factor has taken place in the course of the genetic evolution (phylogeny) of terrestrial plants. However, light also affects the individual development (ontogeny) profoundly insofar as the genes which control normal development of a plant can only express themselves fully in the presence of light. Thus, the normal development of a plant is "photomorphogenesis" (Fig.1).

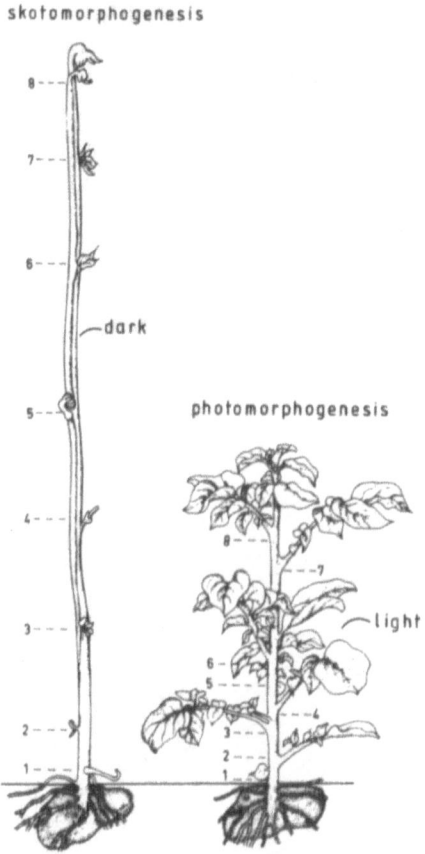

skotomorphogenesis

photomorphogenesis

Fig. 1. Two genetically identical potato plants. Nevertheless, the dark-grown (left) and light-grown (right) plants follow conspicuously different developmental routes ('skotomorphogenesis' is the appropriate strategy of development in darkness, whereas 'photomorphogenesis' is the appropriate strategy of development under conditions of light affluence). The different developmental routes are due to differential gene expressions in light and dark.

The effective light is perceived by a particular sensor pigment, phytochrome (P_{fr}). The study of photomorphogenesis with the means of molecular biology has been particularly attractive since in no other higher organism it is possible to control gene expression by a single and well-defined environmental factor. Even though a vegetative plant is a relatively simple system compared with an 'elephant', its complexity is far beyond the realm of present molecular biology. Therefore reduction of the system has been an essential prerequisite for any study of the 'mechanism' of photomorphogenesis. It turned out [24] that light-mediated plastidogenesis (in the present case the transition of etioplasts to chloroplasts, Fig. 2) is an almost ideal system to model the transition from skotomorphogenesis to photomorphogenesis in the whole plant (see Fig. 1).

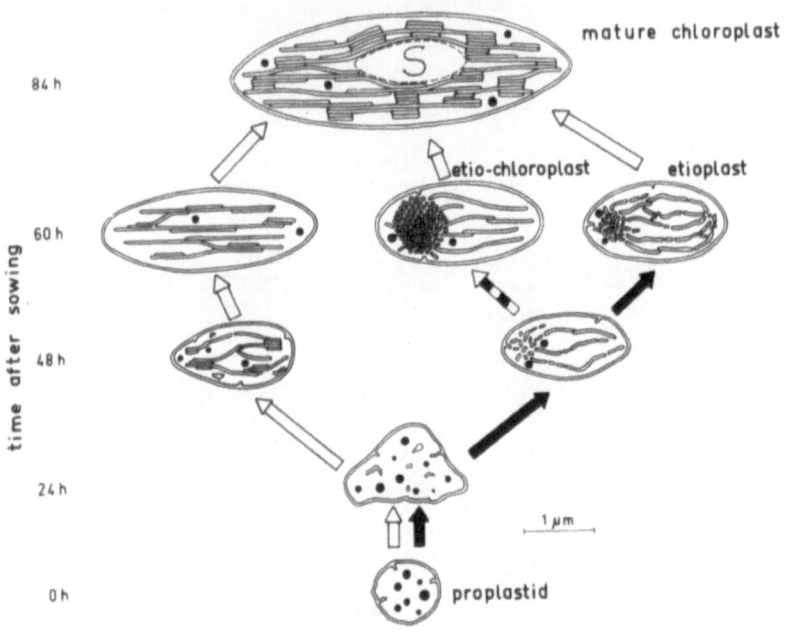

Fig. 2. A scheme showing the time course of plastid development in a typical mesophyll cell in the cotyledons of a mustard plant (representative of higher plants). White arrows, development in continuous white light; dark arrows, development in darkness, leading to an etioplast. Greening and development of a mature chloroplast – characterized by the particular membrane (thylakoid) system – depend on light [26].

Regarding the 'grammar of assembly' (see above) the study of the light-mediated formation of chlorophyll-containing holocomplexes of the photosynthetic membranes (thylakoids, Fig. 2) turned out to be particularly informative. Chlorophyll – synthesized in the light only – does not appear in any free form in the plastid; rather it accumulates as a holocomplex in **stoichiometric** associations with apoproteins and carotenoids. The formation of a holocomplex (5 of them have so far been detected in the thylakoids) shows the essential features of a morphogenetic process since the result is a three-dimensional, highly ordered, unique structure with a characteristic relation between the components (Fig. 3). A coarse regulation of the process is due to the operation of the sensor pigment phytochrome (P_{fr}). Phytochrome determines the rate of synthesis of the components – coarse regulation of a linear production line – at the level of transcription. Another category of regulatory processes – fine tuning – comes into play during assembly of the components. It is this `fine tuning` which guarantees the appearance of a stoichiometrically composed holocomplex. Thus, two layers of control are involved. The first layer is a noisy, rather inaccurate set of regulatory processes – with P_{fr} as the common effector molecule – that generates terminal products of linear production lines as indicated in Fig. 3. The second is a set of refinement processes that corrects for the inaccuracy of the first and yields a precise (and thus functionally reliable) holocomplex.

Components of the antenna complex

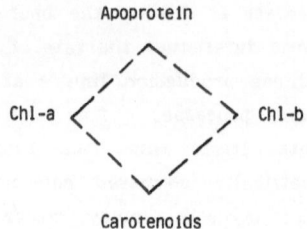

Two modes (layers) of control involved in formation
of antenna complex

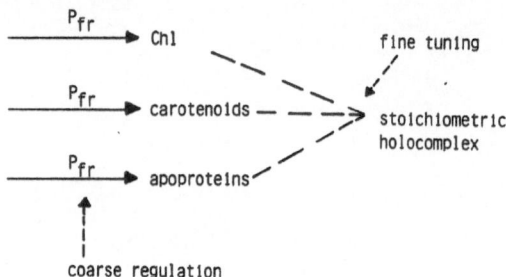

Fig. 3. The formation of the so-called "light harvesting chlorophyll a/b protein complex" of photosystem II (antenna complex) as a model case of holocomplex formation during maturing of the thylakoid membrane (photosynthetic membrane) within a developing chloroplast (see Fig. 2). [23]

In spite of promising achievements of the kind mentioned, the structural problems of development still lie beyond the capacity of any computing facility; it seems that we have to await the solution of the protein-folding problem before the problem of complex formation can be approached in terms of macromolecular chemistry.

This case study tells us, that a holocomplex is encoded in an **distributed** form throughout the genome, and it requires two different types of regulation – one operating at the level of gene transcription, the other operating in connection with molecular assembly – to make a holocomplex. Much the same could be said when we ask, what does it take to make a hand, make a foot, make a liver, make a brain. The specifications for these structures are scattered throughout the genome [5]. To make these structures is not a neat, sequential process, like the linking together of amino acids in a protein. Rather a large number of linear production lines are going on simultaneously, and it is the "grammar of assembly" which determines the specific putting together of the products. To leave no doubt: the ultimate source of information is at the level of the genes, the grammar of assembly is specified in the component parts of the system. As stated above there are at least 5 different holocomplexes. Their specific composition is specified in the nature of the components, in particular the apoproteins.

Clearly, plastidogenesis as a model of morphogenesis has its (narrow?) limits. As a matter of fact, molecular biology of development is the molecular biology of the cell and of the interactions between cells. However, plastidogenesis shows at least that `morphogenesis` is no longer a domain of descriptive developmental history but accessible to the mode of thinking of molecular biology.

On the other hand, sober scrutiny reveals that historical cases of strict reduction in factual science (as contrasted to formal sciences such as logic or mathematics) are extremely rare [1]. The practising scientist, being confronted with the impossibility of strict reduction, is ready to enjoy even incomplete reduction steps such as morphogenesis of holocomplexes. After all, science is the art of the soluble, as Peter Medawar has emphasized, and the analysis of genes and proteins, their biochemistry and their regulation, is – at the time being at least – but a small part of the subject matter of biology. However, despite initial scepticism, issues of cell movement, cellular communication, organ formation, and organisms behavior are rapidly becoming part of the frontier to which molecular biology is contributing [27]. As a final example, use of monoclonal antibody probes to trace the growth and movement of individual identified embryonic neurons to the cells to which they will make connections is a fascinating piece of progress [28].

7. CONCLUDING REMARK (AND PERSONAL CONFESSION)

Is the enforced reduction of as many biological fields as possible – for the sake of a unifying theory of biology – really desirable? I think it is not. To me a consequent and rigorous reduction does not seem to be an urgent task. I even argue against excessive reduction because I am afraid it would impoverish biology. The diversity of biological phenomena requires a diversity of scientific approaches.

The molecular biologist's look at a system cannot be replaced by the comparative biologist's look at the same system, or vice versa. A causal explanation and a functional or teleonomic explanation are complementary rather than mutually exclusive. [10].

Not only is there no reason to eliminate biology as an independent discipline in favour of physics; there is also no reason to reduce the diversity of scientific approaches within biology as long as these approaches are scientifically sound and productive. If, however, a field becomes dominated by rigid paradigms with a concomitant shortage of new ideas, a drive for reduction can stimulate a revolution and concomitant `progress`! Classical and molecular genetics are good examples.

With regard to the `nature` of the living state, I conclude that every process occurring within living systems can (in principle) be explained by a physical mechanism. I further conclude that if we could reproduce experimentally Eigen's [29] stochastic explanation of primordial evolution of biological macromolecules and build up the specific molecular organization of some living state in the test tube, we would have made some kind of a living system. There can be no doubt that this objective is unrealistic at present; however, there is no reason to assume that we will meet any principal obstacles on our way to a man-made living system of some primitive kind. The question is whether this goal is really attractive once we are all convinced that it could be obtained. It might be more worthwhile to elaborate the laws and principles that govern development and behavior of those living systems that have actually been created in huge diversity and utmost sophistication by natural universal evolution.

Bibliography

1. Vollmer, G.: `Reduction and Evolution – Arguments and Examples`, in W. Balzer, D. Pearce and H.J. Schmidt (eds.), **Reduction in Science. Structure, Examples, Philosophical Problems**, Reidel, Dordrecht, 1984.

2. For an excellent, conventional treatment of the problem I refer to Ruse, M: **The Philosophy of Biology**, Hutchinson, London, 1973 and to Hull, D.L.: **Philosophy of Biological Science**, Prentice-Hall, Englewood Cliffs, 1974. Among the philosophers of science, Ernest Nagel has been a strong believer in the possibility of theory-reduction. See Nagel, E.: **The Structure of Science**, Routledge and Kegan Paul, London, 1961.

3. Heisenberg, W.: `Tradition in Science`, in O. Gingerich (ed.), **The Nature of Scientific Discovery**, Smithsonian Inst. Press, Washington D.C., 1975, pp. 219–236.

4. Wuketits, F.M. (ed.): **Concepts and Approaches in Evolutionary Epistemology**, Reidel, Dordrecht, 1984.

5. Lewin, R.: `Why is development so illogical`, **Science** 225 (1984), 1327–1329.

6. Eigen, M. and Schuster, P.: **The Hypercycle**, Springer, Heidelberg, 1979.

7. Küppers, B.O.: `Das Paradoxon der Evolution`, in B. Kanitscheider (Hrsg.), **Moderne Naturphilosphie**, Könighausen und Neumann, Würzburg, 1984, pp. 317–335.

8. Medawar, P.B. and Medawar, J.S.: `Reductionism`, in **A Philosophical Dictionary of Biology**, Harvard University Press, Cambridge, Mass., 1983, pp. 227–232.

9. Mohr, H.: `Freiheit und die biologische Natur des Menschen`, in P. Koslowski, Ph. Kreuzer und R. Löw (Hrsg.), **Evolution und Freiheit**, Civitas Resultate Band 5, Hirzel, Stuttgart, 1984, pp. 36–53.

10. Mohr, H.: **Lectures on Structure and Significance of Science**, Springer, New York, 1977.

11. Woodger, J.H.: **Biology and Language**, Cambridge University Press, Cambridge, 1952.

12. Watson, J.D. and Crick, F.H.C.: `A structure for desoxyribose nucleic acid`, **Nature** 171 (1953), 737–738.

Watson, J.D. and Crick, F.H.C.: `Genetical implications of the structure of desoxyribonucleic acid`, **Nature** 171 (1953), 964–967.

13. Jacob, F. and Monod, J.: `Genetic regulatory mechanisms in the synthesis of proteins`, J. Mol. Biol. 3 (1961), 318–356.

14. Polanyi, M.: `Life's irreducible structure`, Science 160 (1968), 1308–1312.

15. Maddox, J.: `Is Biology now Part of Physics`, Nature 306 (1983), 311.

16. Mothes, K. (Hrsg.): **Biologische Modelle**. Nova Acta Leopoldina, No. 184, Vol. 33, Deutsche Akademie der Naturforscher Leopoldina, Halle, 1968.

17. Hess, B., Boiteux, A. and Kuschmitz, D.: `Regulation of Glycolysis`, in H. Sund und V. Ullrich (eds.), **Biological Oxidations**, 34. Colloquium – Mosbach. Springer, Heidelberg, 1983, pp. 249–266.

18. Estes, W.K.: `Human Behaviour in mathematical Perspective`, Amer. Scientist 63 (1975), 649–655.

19. Rychlak, J.F.: **A Philosophy of Science for Personality Theory**, Mifflin, Boston, 1968.

20. Quoted from Stent, G.S.: `Molecular Biology and Metaphysics`, Nature 248 (1974), 779–781.

21. Mohr, H. und Sitte, P.: **Molekulare Grundlagen der Entwicklung**, BLV, München, 1971.

22. Mohr, H.: `Pattern Specification and Realization in Photomorphogenesis`, in W. Shropshire and H. Mohr (eds.), **Photomorphogenesis**, Encycl. Plant Physiol., Vol. 16A, Springer, Heidelberg, 1983, pp. 336–357.

23. Bryant, S.V., French, V. and Bryant, P.J.: `Distal Regeneration and Symmetry`, **Science** 212 (1981), 993–1002.

24. Kasemir, H. and Mohr, H.: `Chloroplastenentwicklung – ein Modell für Morphogenese`, Biodidactica (1985).

25. Mohr, H. and Shropshire, W.: `An Introduction to Photomorphogenesis for the general Reader`, in W. Shropshire and H. Mohr (eds.), **Photomorphogenesis**, Encycl. Plant Physiol., Vol. 16A, Springer, Heidelberg, 1983, pp. 24–38.

26. Mohr, H.: `Phytochrome and Chloroplast Development, in N.R. Baker and J. Barber (eds.), **Chloroplast Biogenesis**, Elsevier, Amsterdam, 1984, pp. 306–347.

27. Blattner, F.R.: `Biological Frontiers`, **Science** 222 (1983), 719–720.

28. Goodman, C.S. and Bastiani, J.: `How Embryonic Nerve Cells recognize one Another`, Sci. American, 251, No. 6, December 1984, pp. 50–58.

29. Eigen, M.: `Selforganization of Matter and the Evolution of biological Macromolecules`, **Naturwissenschaften** 58 (1971), 465–523.

Günter P. Wagner

THE VARIANCE ALLOCATION HYPOTHESIS OF STASIS AND
PUNCTUATION

INTRODUCTION

Reductionistic research strategies are a powerful heuristic instrument in natural sciences provided that their inherent limitations are not obscured by ontological reductionism (Wuketits 1983). Reduction of complex phenomena to basic and more or less simple processes is feasible if one of two prerequisites is realized. Either the more complex phenomenon can be traced back to a number of more elementary processes among which only weak interactions occur. The paradigm of this type of reduction is statistical gas theory at supercritical temperature and at low pressure, or population genetic theory of large populations with weak selection and the absence of epistatic interactions. Both, the gas molecule as well as the gene can be treated independently of the rest of the system. The other situation where reduction is feasible, is given if there exists a simple linear superposition rule which allows to predict systems behavior from that of elementary processes. An example is the interaction of electromagnetic waves at low amplitude, i.e. linear optics. In any other case the study of elementary processes has to be complemented by a kind of systems theory.

Systems theory, as the term is used here, is a tool to deal with complex situations. It can be useful even in cases which one could deal without. However, there are problems in which systems theory is stringently required. Such problems occur for instance, if the elements of a system are cyclically connected as in a regulatory system. Well known parts of systems theory are linear systems theory from electrical engineering, cybernetics, and synergetics, which deals with cooperative effects in systems with nonlinear interactions (Haken 1977).

161

P. Hoyningen-Huene and F. M. Wuketits (eds.), Reductionism and Systems Theory in the Life Sciences, 161–185.
© *1989 by Kluwer Academic Publishers.*

However, holistic or systems theoretic approaches are only needed if reductionistic approaches have failed. Purely philosophical justifications of holistic approaches are of little help in natural sciences. Hence, the first step within a systems theoretical research program is to identify the reasons why a reductionistic approach has failed in the particular case. This question is raised in the present paper with reference to the recent debate on the explanation of macroevolutionary phenomena.

In evolutionary theory, the reductionistic research strategy has been very successful during the last fifty to seventy years, culminating in the population biologic foundation of evolutionary theory (Mayr 1982). Only recently some serious problems concerning the explanatory power of population genetic theory became evident: partly stimulated by the punctualist–gradualist debate, partly by the rush of developmental genetics (e.g. what is the evolutionary role of homeotic genes?).

In the present paper I want to discuss some of the problems of the punctualist–gradualist debate from a systems theoretic perspective. Exactly which of the aspects I would like to discuss is explained in section 2.

In order to avoid confusion regarding the perspective of a systems theory of evolution (subsequently called STE) a short outline of what STE is about and how STE is related to population biological theory of evolution will be of help (see also Wagner 1983, 1985, 1986).

STE is an attempt to study the peculiarities of organismic evolution on the basis of known evolutionary mechanisms. No other elementary mechanisms of evolutionary change are assumed to exist than those already described in neo–Darwinian theory: mutation, recombination, selection, drift, meiotic drive, isolation, and so on. In addition it is asked what factors can account for the observed regularities in the overall pattern of morphological evolution, e.g. stasis, the origin of new body plans, the stability of design within a lineage. These factors cannot be, by the virtue of the first assumption, new unknown mechanisms of evolution. Instead in STE it is asked how the boundary conditions of organismic evolution may modify the action of known evolutionary mechanisms. More precisely, the question is, what boundary conditions can last long enough to explain long term trends in morphological evolution.

It has to be recalled that the evolution of any lineage is a unique historical event. As such, regularities observed in the evolution of different lineages cannot be explained by some general natural law comparable to the laws of physics, because the known mechanisms of evolution leave too much room for contingencies. The only possibility to explain regularities in any historical development is to look for boundary conditions that last long enough to induce the observed regularities. This problem appears to be the same in all sciences dealing with historical entities (for a lucent discussion of this problem see Popper 1979).

There are many levels of organization at which long lasting boundary conditions of the evolution of a lineage can emerge. On the ecosystem level by interspecific interactions (e.g. the Red Queen hypothesis, Van Valen 1973, Stenseth and Maynard Smith 1984), at the organismic level by the materials utilized during development (Alberch 1980) or by fabricational processes (Seilacher 1973). There is little doubt that all these factors play an important role in delimiting the route of evolutionary change, but STE chooses an approach that is not covered by the before mentioned theories.

STE concerns those boundary conditions of organismic evolution that emerge from the formal properties of the evolutionary mechanism themselves. Boundary conditions of organismic evolution of this kind emerge because the organism is an integrated system and the primary object of natural selection. STE thus concerns system properties that can be summarized under headings like adaptability, complexity, and so on (see below).

At the most general level STE can be considered as the study of the formal properties of large aggregates of characters or genes and their evolutionary dynamics. It is thus comparable to cybernetics, where the boundary conditions for the structure of regulatory systems are derived from general formal properties of feedback cycles. The principles of cybernetics apply to the human locomotory control system as well as to steam-engines.

In this sense STE is a direct expansion of the neo-Darwinian research program. However, it is intended to analyze how evolutionary dynamics of organisms may differ from the predictions obtained from single character models. It is expected that these formal properties of organismic systems may account for general features of morphological evolution, such as the mosaic mode of evolution, irreversibility, or the episodic pattern of morphological evolution.

Three aspects of the evolution of organismic systems have been considered in the last years:

i) **Reflexivity of adaptation** denotes the fact that a character has to adapt not only to the extraorganismal environment but also to the other characters within the same organism. The same principles that hold for the adaptation of characters to the external environment need also to be true for the coadaptation between the characters of one organism. It can be shown rigorously that reflexivity is sufficient to explain the irreversibility of

morphological evolution (Wagner 1982) and may account for the fact that
macroevolution is not controlled entirely by environmental factors (Wagner
1983, 1985).

ii) For the **adaptability** of a set of quantitative characters the existence of
additive genetic variation is not sufficient. This implies a number of problems
regarding the evolution of functionally coupled characters (Bürger 1985,
Wagner 1984a, 1988a,b). We will return to this problem in section 4.1.

iii) The impact of **phenotypic complexity** on the stability of the population. How
many stochastically independent characters can be controlled by stabilizing
selection simultaneously?

In summary, STE can be characterized as the study of how the evolution of a
character is influenced by the fact that there are other characters parts of the
same organism.

It will be shown that apparent alternatives in the explanation of stasis and
punctuation, namely developmental constraints and stabilizing selection or random
speciation and adaptation, are complementary rather than exclusive. Both the
theory of optimizing selection as well as the population genetic theory of
adaptability leads to the conclusion that the variation of the phenotype has to be
constrained in some respect in order to allow natural selection to be effective.
Hence, the existence of constraints has to be predicted on the basis of
population-genetic theory.

2. PHENOMENOLOGY OF MORPHOLOGICAL EVOLUTION: GRADUAL OR EPISODIC TRANSFORMATION?

The term "episodic transformation" is used in this paper in order to decouple the
subsequent discussion from the main lines of the punctualist-gradualist debate. It
is concerned with an aspect of morphological evolution that has some affinity to
the punctualist mode of evolution. However, it is not yet clear whether all parts
of the punctuated equilibrium model are true. The theory of punctuated equilibria
is a system of hypotheses including propositions about the time course of
morphological evolution, the mode of speciation, and the mechanisms of
macroevolution (Eldredge and Gould 1972).

While there is little doubt that some characters display episodes of rapid
evolutionary transformation, separated by phases of little change, there is
considerable disagreement concerning the causes of this mode of phenotypic

evolution. According to Gould and Eldredge, the reason is a coincidence of morphological change and allopatric speciation. Indeed, there are examples where a sudden change in morphology is associated with a short interval of parent-daughter species overlap in the fossil record, suggesting allopatric speciation and subsequent replacement of the mother species. Examples are the transition between the Jurassic brachiopodes *Kutchithyris acutiplicata* and *K. euryptycha* (Ovcharenko 1969) or the reduction of the metaconid occurring at the transition between *Oxyaena forcipata* and *O. ultima* in early Eocene (Bakker 1983).

However, there are also instances where the fossil record suggests a rapid shift in morphology occurring in widespread populations simultaneously. Malmgren et al. (1983) demonstrated a sudden phyletic change of shape in the planktonic formaminifera genus *Globorotalia* at the Miocene–Pliocene boundary. The transition from *G. plesiotumida* to *G. tumida* can be observed in most Miocene–Pliocene boundary sections from Indo–Pacific equatorial regions. Even proponents of punctuated equilibria provide data at variance with the allopatric speciation mode. The famous punctuational transformation of fresh water molluscs observable in the sediments of lake Turkana basin, indicates that the evolutionary changes at the Surgei level, although rapid, occurred over a large area and in thick faunal units containing many millions of individuals (Williamson 1981). This observation together with the fact that the obligatory asexual species *Melanoides tuberculatus* displays also the punctual mode of evolution questions the significance of genetic drift, founder effects and inbreeding as the only mechanisms of punctuation.

The hypothesis of a direct and necessary association between speciation and morphological change is falsified by the evidence available on the salamander genus *Plethodon* are highly differentiated in genetical terms (Wake 1981, Wake et al. 1983).

Indirect tests on the association between speciation and the rate of morphological evolution lead to equivocal results (Douglas and Avise 1982, Schopf 1982, Cherry et al. 1982). Thus punctuation in morphological evolution may not necessarily be coupled with allopatric speciation. Therefore punctuation and allopatric speciation should be discussed separately. In order to emphasize the difference between punctuation in the strict sense of the word from the phenomenon of rapid morphological change the term "episodic transformation" will be used.

2.1. THE OLENUS STORY

Different modes of evolution are usually explained by examples showing only punctuation or phyletic transformation. In either case a variety of arguments is at hand explaining how the observed mode of evolution can be the consequence of artefacts. In no instance it is possible to exclude all pitfalls imaginable. In this respect the evolution of the Cambrian trilobit genus *Olenus* is a favourable exception. Kaufmann (1933) was able to demonstrate both modes of evolution in a single lineage. The different modes occur in the evolution of different characters. A similar pattern was observed in other trilobits (the Devonian Genus *Phacops*) by Eldredge (1971). Rather complicated scenarios are necessary to explain this pattern of evolution as a result of artefacts. Because the paper of Kaufmann is not available in English a more detailed review is given below.

The stratigraphic column described by Kaufmann (1933) is about 37 m thick and involved five zones containing fossil *Olenus* species. Within the five zones six species were found: *Olenus gibbosus, -transversus, -truncatus, -Wahlenbergi, -attenuatus,* and *-dentatus.* Evolution is mainly described on the basis of head and pygidial characters. Three zones were thick enough to display substantial phyletic evolution in four species: *Ol. gibbosus, -truncatus, -attenuatus,* and *-dentatus.* The latter two species occur in one continuous zone and show overlap in a narrow band, suggesting a parent-daughter species displacement.

In all phyletic series an iterative trend in pleura reduction was found at the heads as well as at the pygidial sections. In all series the relative width of the glabella and the rhachis remains constant, i.e. display stasis. A phyletic trend in eye-fillet length was found in *Ol. attenuatus* and *Ol. dentatus.* No apparent trend of eye-fillet evolution was seen in the series of *Ol. gibbosus.* The trends in *Ol. attenuatus* has opposite direction as in *Ol. dentatus,* even if *Ol. dentatus* most probably is a direct descendant of *Ol. attenuatus.*

Characters that display stasis within species are nevertheless variable between species. The glabella becomes wider in ascending series, while the rhachis varies without obvious trend. Some characters are variable only among a subset of species. For instance the pygidial spines become shorter between the *Ol. transversus* and *Ol. truncatus.* A sudden change in the phyletic trend of eye-fillet evolution was observed at the transition from *Ol. attenuatus* to *Ol. dentatus.* This change in phyletic trend is accompanied by a change in ontogenetic allometry of the eye-fillet.

In summary, within a series of species from one genus there are characters that show regular phyletic transformation series while other characters are constant within each species but vary instantaneously between species.

2.2. EXAMPLES OF GRADUAL PHYLETIC EVOLUTION

Most of the examples of gradual phyletic evolution comprise data of tooth size evolution in mammals (Kurtén 1959, Gingerich 1974, Bookstein et al. 1978). Regular oscillations of tooth size during Pleistocene are known to correlate with glacial and interglacial periods, suggesting an environmentally driven adaptive modification. Also in Eocene longlasting and iterative phyletic trends of tooth size are apparent in *Hyopsodus* (Gingerich 1974). In this example species with divergent size trends occur sympatrically, suggesting that speciation has preceded the change in size.

But gradual phyletic evolution in tooth characters is not restricted to change in overall size. Rose and Bown (1984) have shown that the transition between the omomyid *Tetonius honunculus* and *Pseudotetonius ambiguous* in early Eocene is achieved by a gradual phyletic diminution of the teeth between I_2 and P_4 lasting at least for 4.10^6 years. The very much reduced P_2 of *Tetonius homunculus* in the first stage of the series indicates that the trend most probably has started before the transition between *Tetonius* and *Pseudotetonius* began.

Hominid fossils are sometimes used to demonstrate stasis (e.g. Rightmire 1981) or gradual phyletic evolution (Croin et al. 1981). However, as long as the phylogeny of major hominid taxa remains unclear, inferences about the mode of evolution are impossible (Arthur 1984).

Rigorous statistical tests of the existence of directed gradual modification in the evolution of the Jurassic ammonite *Zugokosmoceras* led to equivocal results (Raup and Crick 1981). Only one parameter (ratio of umbilical diameter with shell diameter) shows directional change if sampled over 100cm intervals. The evolution of shell diameter is consistent with a random walk model, but under special sampling conditions also with directional change.

Although the overall patterns of Equidae evolution is not orthogenetical, there are some trends in the transformation of the skull proportions. This fact was interpreted by Robb (1935a,b) as a simple allometric consequence of size increase. This question was recently reinvestigated by Radinsky (1984). Radinsky found a mosaic of a general allometric trend extending over the whole horse evolution, intermitted by one change in skull proportions between *Mesohippus* and

Merychippus. This nonallometric change regards the position of the cheek tooth row relative to the orbit and the position of the occlusion plane relative to the mandibular– squamosal joint. This transformation most probably is related to the increase in crown height at the origin of *Merychippus.* Nevertheless already *Merychippus* has the skull proportion of *Equus,* if the size related allometry is taken into account, even if relative tooth height in *Merychippus* is only 60 % that of *Equus.* Thus the reorganization of skull proportions appears to have occurred in a threshold manner.

2.3. CASES OF EPISODIC TRANSFORMATION

The discussion of horse skull evolution (see above) gave some indication of the existence of instantaneous changes interspersed in a broad pattern of harmonic allometric evolution, even if the time course of the event is not described. Besides the already mentioned examples (shape in foraminifera, Malmgren et al. 1984, shape of freshwater molluscs, Williamson 1981, and metaconid evolution in *Oxyaena,* Bakker 1983) the most convincing data on stasis and episodic change concern the coevolution of mammalian cursorial predators and their ungulate prey (Bakker 1983).

Coevolution of ungulates and their cursorial mammalian predators provide an exceptionally instructive illustration of how morphological evolution precedes. The reasons are: 1. The functional anatomy of quadrupedal running is well understood and can be reconstructed from the fossil remains of ungulates and carnivores, 2. there is an iterative evolution of large cursorial predators (Mesonychids, Hyaenodonts, Amphicynodonts and Canids), and 3. the life table of extinct ungulates can be reconstructed from the frequency distribution of tooth wear classes.

The degree of locomotory adaptation of the appendicular skeleton was described mainly on the basis of two traits: 1. The metatarsal–femur index and 2. the grooving of the astragular–tibial joint. Both characters show no indication of gradual modification within the carnivore lineage (Bakker 1983). Evolution, however, does occur into the predicted direction at the transitions between species, but stasis can last up to 3.10^6 years as for instance in the mesonychid species *Pachyaena gracilis.* This species became replaced by a more advanced type in late Eocene indicating the possibility to attain a higher locomotory adaptation grade within the mesonychid design. Stasis of suboptimal phenotypes is even more pronounced in the Hyaenodonts which replaced the mesonychidae **after** they perished at the Eo–Oligocene boundary. From Eocene to Oligocene the ungulates

have further improved their locomotory adaptations such that the adaptive lag
between predators and ungulates in Oligocene is much wider than in late Eocene.
The most advanced hyaenodon (*Neohyaenodon*) in mid Oligocene represents an
adaptive grade inferior of mid Eocene mesonychid *Dromocyon*. This is confirmed by
the frequency distribution of tooth wear classes of ungulates from Oligocene as
compared to ungulates from Eocene. These curves strongly suggest that Eocene
predators had less difficulty pursuing and catching young, healthy adult ungulates
than did Oligocene predators.

While the direction of evolution is very well predictable from functional
considerations, the rate and episodic mode of evolution is less well understandable
on the basis of known genetic mechanisms. Stasis can last for many generations
even if later species show, that improvement was possible in principle.

A narrow zone of overlap of two successive species is usually taken as evidence
for punctuation caused by allopatric speciation (see above Ovcharenko 1969,
Bakker 1983 regarding tooth-evolution in Oxyaenids). More complete evidence from
a broad paleogeographic survey is given for the Devonian trilobit species *Phacops
rana* by Eldredge (1971). Within the species *Phacops rana* phyletic as well as
episodic transformations are found. Among the characters that show episodic
changes, the evolution of the number of eyes alined in the dorso-ventral columns
of the eyefield are best documented. There is a trend to reduce the number of
eyes per column from 18 to 15, which is the major evolutionary change in *P. rana*
stock. The new phenotypes occur first in populations marginal to the range of
distribution followed by a subsequent replacement of the old phenotype over the
range of the species.

From this example it becomes evident that episodic evolution also occurs in
meristic characters and can be caused by allopatric speciation.

2.4.CONCLUSIONS

By comparing the various case histories described in sections 2.1, 2.2 and 2.3 it is
evident that a particular mode of evolution is not consistently associated with a
certain kind of character. There is some prevalence of size related changes among
the examples of gradual phyletic evolution, but a sudden dwarfing is also possible,
for instance in mammals at the end of the Würm glacial age (Kurtén 1959). Other
quantitative characters, for instance the proportion of the metatarsus relative to
the femur, evolve episodically in mammalian cursorial predators (Bakker 1983).
Meristic characters can evolve either episodically (eye-number in *Phacops rana*,

Eldregde 1971) or gradually (enamel ridges in the cheek teeth of elephants, Maglio 1973). Moreover, the mode of evolution may change from stasis to phyletic evolution as in the genus *Olenus* (Kaufmann 1933). The mode of evolution is thus not characteristic of certain types of characters.

A simple variation in selection intensity is not sufficient to account for the different modes of evolution. Episodical transformations may be triggered by a major environmental perturbation as at the Surgei level in the Turkana basin sediments (Williamson 1981). On the other hand stasis can persist even if the current phenotype is obviously suboptimal, as suggested by the efficiency of predation and the morphology of species which originated later (Bakker 1983). Hence, the mode of evolution can not always be explained by variations in selection intensities even if severe environmental perturbations may be able to trigger episodic transformations.

The only general conclusion derivable from the above examples is that in any species there is only a limited number of characters that undergo gradual phyletic evolution simultaneously. Stasis of the remaining characters can be overridden under certain circumstances, possibly during allopatric speciation and under hard selection pressure. The pattern of gradually modifiable characters and characters showing stasis can change, and may be characteristic of a lineage.

In this context it may be worth to mention some observations well known to paleontologists: the phenomenon of mosaic evolution. It refers to the fact that the characters contributing to a certain body design have not evolved in concert. Their rate may vary within the same lineage at a given time. One or more of the characters evolve first, and achieve their final stage before other characters begin to change.

An excellent demonstration of the erratic nature of mosaic evolution is possible if parallel evolution has occurred, as in the evolution of horses. The tooth and limb characters of *Equus* originated with greater temporal lag than necessary. This can be concluded from the parallel evolution of south American ungulated during the Oligocene and early Miocene (Notohippidae and Proterotheriidae). Hysodont teeth and one toed limbs were achieved much earlier in these and other lineages (e.g. *Plagiolophus* during Eocene) than in the Equid family (see Simpson 1951). Furthermore the tooth and limb characters evolved in different lineages independently, indicating the lack of any necessary interaction between the two functional specializations. It was only a matter of chance and not of adaptive necessity that the Equidae obtained both features.

Similarly bipedal locomotion was achieved at least 1–2 million years earlier in the *Australopithecus* lineage than the *Homo* lineage started with brain enlargement (Johanson and White 1979). Mosaic evolution disproves all theories that postulate a direct dynamic coupling between the origin of bipedal running, use of tools and brain enlargement (e.g. Washburn 1960). There is only a permissive or conditional relationship between the evolution of these characters.

In summary, the pattern of morphological evolution is characterized by an interdigitation of adaptively versatile characters with a highly constrained background. Adaptively versatile characters may eventually show gradual phyletic evolution. The complex of constrained traits episodically releases some characters, subsequently allowing gradual adaptive evolution. Eventually, parts of the constrained character set become suddenly transformed under certain circumstances but remain in stasis after punctuation.

The remaining sections of this paper are devoted to the question whether there is any evolutionary principle that makes a highly constrained design of the organisms necessary. It will be argued that the population genetic consequences of phenotypic complexity may be the reason for the described pattern of evolution.

3. THE VARIANCE ALLOCATION HYPOTHESIS

From the pattern of morphological evolution an image of the organism emerges that is dominated by an interplay between adaptability of some characters and constraints in different parts of the body. During evolution, adaptability may get lost for some characters and constraints may be broken. Hence, evolution is to a great extend, at least at the interspecific level, the history of changing patterns of adaptive versatility and constraints. It is proposed here that this aspect of organismic evolution is best described at the level of variance allocation and not so much on the level of developmental mechanisms. It is proposed that morphological evolution should be studied at two levels: 1. the adaptation of characters or character combinations currently accessible to natural selection, and 2. the change of the pattern of constraints and versatility.

The term **variance allocation** is used in analogy to the term energy allocation. Energy allocation refers to the problem how to use a limited amount of energy for growth and reproduction such that fitness is maximized. The hypothesis of variance allocation assumes that only a limited amount of total variance of phenotypic characters is compatible with adaptability and stability of populations. Arguments supporting this assumption are presented in section 5. If this is true the problem occurs of how to allocate this limited amount of variance to a limited set of characters such that the probability to go extinct is minimized and the chance to cope with environmental perturbations by genetic adaptation is maximized.

The variance allocation hypothesis suggests that the interplay between phyletic adaptation, constraints, and episodic transformations is a consequence of the optimization problem resulting from the limited amount of flexibility allowed in the design of organisms. Which mechanism are able to change variance allocation is an open problem.

3.1. NATURAL HISTORY OF CONSTRAINTS

"Constraints are features of ontogenetic mechanisms and morphogenetic design which limit the power of selection to mold phenotypic traits" (Stearns 1982, p. 239). The concepts used to illustrate the role of constraints vary between basic physicochemical laws, the formal properties of pattern formation processes, fabricational limitations of design and pleiotropic gene actions. Which view of constraints is preferred depends on the opinion of the author regarding the degree of their immutability and their supposed role in organismic evolution. The range of opinions extends from eternal importance of constraints up to complete modifiability leaving no substantial problem for natural selection to work (Charlesworth et al. 1982).

In this paper a compromise is suggested. Clearly, constraints exist, are not modifiable at will, but may change during evolution.

An analysis of covariance among life history traits in mammals performed by Stearns (1983) indicates lineage specificity of constraints, e.g. there is a much higher degree of integration found in felids than in canids even though both belong to the same order. A large amount of total covariance is caused by variation among higher taxonomic units (e.g. families) indicating two things: 1. The variation among higher taxa is possibly constrained by some general functional demands, causing the high contribution of interfamily covariation to total covariation. 2. The covariation within families varies substantially from family to family, showing lineage dependency.

Lineage dependency of constraints may have several reasons: 1. Contingencies due to the origin of the particular lineage and their prevailing mode of life, 2. modification of the system of constraints in the course of evolution within each lineage.

An example most probably illustrating a contingent difference in developmental constraints of Anurans and Urodeles was described by Alberch and Gale (1983). Dwarfing lineages in amphibians generally show a tendency to reduce the number of digits. Which digit is lost depends on the lineage. This phenomenon can be reproduced experimentally by inhibiting mitosis in early limb bud development. Anurans preferably lost phalange from toe 1 while urodeles preferably from toe 5. Clearly there is a lineage dependent constraint of morphological variation caused by differences in the timing of the developmental process.

Another approach to demonstrate the lineage dependency of adaptive versatility and constraints is the study of the phylogeny of specific organs. The phylogeny of the gastropod shell provides an example of increasing adaptive versatility during evolution (Vermeij 1973). The number of parameters necessary to characterize shell form increases from two in the early Cambrian uncoild sells up to six of the conispirally coiled forms in the late Cambrian times. Vermeij noted that increased versatility of shell forms led to an increase in the number of possible solutions to any given mechanical problem (Vermeij 1973).

Well known to comparative anatomists is also the observation that the evolution of higher organisms not only proceeds by increasing adaptive versatility in structure but also by acquisition of new constraints. Acquisition of new constraints is usually recognized by taxonomists as new diagnostic features defining higher taxonomic units. Most convincing are examples where the number of meristic characters is fixed in the more advanced taxon and variable in the taxons closer to the ancestral condition. All of the over 18000 species of the crustacean subclass Malacostraca have 8 thoracal segments while within the Brachiopod suborder the number of thoracal segments varies between 10 and 32. The number of neck vertebra is fixed in mammals but variable e.g. in the reptilian orders Plesiosauria and Therapsidae. In fact the whole phylogeny of vertebrates can be considered as the history of newly acquired constraints (Riedl 1979).

The examples mentioned above show that it may not be wise to stick too closely to a certain material explanation of constraints. Macroevolution proceeds in overcoming the constraint of earlier stages of evolution and acquiring new constraints at higher organizational levels.

3.2. STATISTICAL THEORY OF VARIANCE ALLOCATION

Variance allocation is most conveniently described in the case of continuously varying characters. Provided that the characters are normally distributed the correlation matrix is sufficient to characterize the allocation of variance in state-space. A number of statistical techniques are at hand to analyse the structure of correlation matrices such as factor analysis, principal component analysis (PCA), multiple regression analysis, and cluster analysis. Among them principal component analysis is of special importance as their statistical meaning is obvious.

In PCA the eigenvalues and eigenvectors of the correlation matrix is extracted. The eigenvalues are proportional to the amount of variance dispersed along the subspace of the corresponding eigenvector. Therefore the distribution of eigenvalues reflects the distribution of variance among the dimensions of the state space. If only few eigenvalues are large and the remaining ones close to zero then the character dispersion is highly restricted to a few dimensions of the state space. On the other hand, if all eigenvalues have more or less the same magnitude then the dispersion is nearly symmetrical, each dimension in the state space has approximately the same share of total variance. The higher the variance of the eigenvalues the more restricted is character variation (Wagner 1984b).

A concept closely related to variance allocation is morphological integration (Olson and Miller 1958). It refers to the observation that characters related in function or development tend to be more tightly correlated than other characters. This concept has been extended to the structure of genetical correlation matrices revealing similar patterns of association (Cheverud 1982).

In order for the variance allocation hypothesis to be sensible the degree of organization reflected in genetic and phenotypic correlation matrices should be higher than expected by chance. According to random matrix theory the eigenvalues of a correlation matrix of randomly associated characters should be exponentially distributed with an variance slightly less than 1 (Wagner 1984b). Actually all genetic and phenotypic correlation matrices of mammalian skeletal characters have an eigenvalue variance much greater than 1. In contrast, correlations of characters not as closely related in function as skeletal characters have an eigenvalue variance of approximately one.

A high degree of integration is maintained among functionally related characters and this degree of integration is higher than predicted by chance as presupposed for the variance allocation hypothesis.

4. POPULATION GENETIC THEORY OF COMPLEX PHENOTYPES

In this section some results are reported that indicate that the relationship between developmental constraints and evolution by Darwinian selection is more complicated than previously thought. It can be shown that constraints not only can inhibit adaptation but may be a prerequisite of adaptation in certain situations. Adaptation by natural selection may be impossible without complementation by constraints (see 4.2.).

4.1 THE INFLUENCE OF DEVELOPMENTAL CONSTRAINTS ON THE RATE OF MULTIVARIATE PHENOTYPIC EVOLUTION

Intuitive reasoning about tempo and mode of evolution is usually based on the assumption that the rate of evolution is a monotonically increasing function of selection intensity and the amount of additive genetic variation. As a corollary, it is often concluded that stasis or bradytelic evolution has to be either due to the absence of directional selection (i.e. stabilising selection) or to the lack of genetic variance on which selection can act. However, quantitative genetic theory shows that the rate of evolution is a systems property, depending not only on the variance of those characters currently under directional selection, but also on the total pattern of variation of the whole phenotype. This fact is especially relevant for the evolution of functionally interdependent characters (Wagner 1988a,b).

The evolutionary dynamics of functionally constrained characters can be modelled by so-called "corridor models" (Rechenberg 1973, Wagner 1984a). A corridor model is an adaptive landscape looking like a ridge, i.e. a fitness function $m(z)$ $(z=(z_1,...,z_n)$, z_i a quantitative polygenic character), where there is a "path" along which directional selection occurs and stabilizing selection is acting in all directions orthogonal to it. In adaptive landscapes of this type rather peculiar phenomena can occur, one of them may be called the "evolution window". If one analyses, on the basis of quantitative genetic theory, how an increase in variance of all characters influences the rate of evolution (under constant heritability), one realizes that the rate of evolution can decrease if total variance exceeds a certain level (Fig. 1).

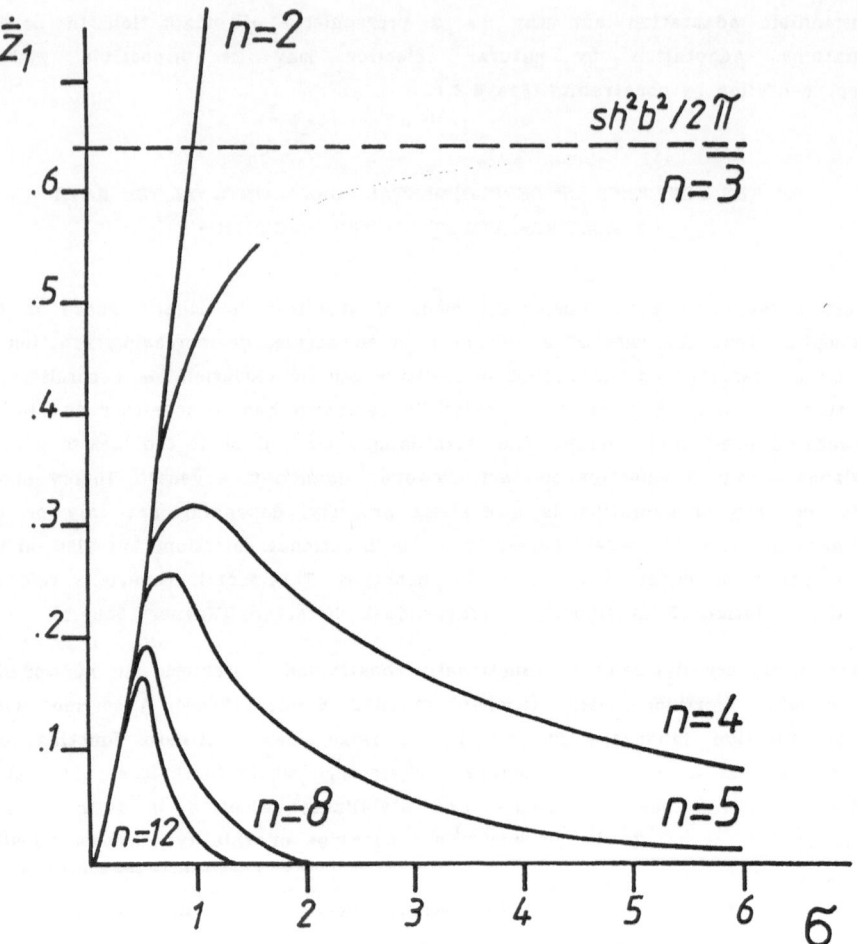

Fig.1: Influence of variation q of characters (heritability = 1) on the rate of phenotypic evolution za given the fitnessfunction m(z)=(sz₁, if |z₁| ≤ b/2 for all i≠1, and −q, if there is at least one i≠1 with |z₁| > b/2). The number of characters is n, b=2 and s=1. Note that the rate of evolution can decrease if variation of characters exceeds a certain optimal level.

The term "evolution window" refers to the observation that evolution can only proceed if the amount of total variance is within certain limits. In addition, the structure of the phenotypic covariance matrix is of critical importance (Bürger 1985).

The reason for this phenomenon is that variation along the direction of the ridge has an effect on the rate of evolution opposite to that of variation orthogonal to it. Increased variation along the line of directional selection accelerates evolution, but variation orthogonal to the direction of evolution ("lateral variation") can **inhibit** evolution. This phenomenon occurs in a large class of adaptive landscapes but not in all. The inhibitory effect never occurs for instance in adaptive landscapes of functionally unconstrained characters which contain only one optimum, but is found in so called "malignent" fitness functions (Bürger 1985, Wagner 1984a, 1988a,b,c). Malignent fitness functions are models of functionally interdependent characters (see Wagner 1988a,b). These results suggest that the rate of evolution of functionally integrated phenotypes can be influenced very much by the pattern of **phenotypic** variation (in addition to the amount of additive genetic variation under directional selection). Restrictions in the amount of "lateral variation" of functionally coupled characters can be of adaptive significance. Constraints can accelerate the rate of phenotypic evolution if their pattern corresponds to the system of functional constraints.

Functional interdependences of the type that lead to malignant fitness functions can be expected to occur between the vegetative organs and the rest of the body. For instance, any improvement of the locomotory system will only be correlated with higher fitness values if the cardio-vascular and the respiratory system is well functioning. The same is true of the liver and the excretory system. Hence, maladaptive variation of these organs will deflate the correlation between locomotory performance and fitness. Adaptation of the locomotory system by directed natural selection requires that the variation of the vegetative system is kept within certain limits to avoid deflation of selection intensity.

Characteristic of the growth and development of these organs is that it is strongly coupled to the demand of the skeletal like characters (CNS; locomotory system) by epigenetic regulation of their growth. This type of growth regulation can be called liverlike, because this kind of regulation was first found in the liver (Stark 1975).

In contrast, the growth regulation of skeletal like characters is largely autonomous and more or less under direct genetic control. The adaptive significance of these character is primarily found in adaptation to the environment. This difference in the biological role of liver like and skeletal like characters (Wagner 1988c), is well in agreement with the expectations from the theory of adaptability. Organs which provide the functional prerequisites for the adaptive significance of environment related characters avoid independent variation by a strond epigenetic coupling of their growth to the demands of the skeletal like characters.

4.2. STABILIZING SELECTION AS A CAUSE OF STASIS

To geneticists stabilizing selection appears to be the most plausible cause of stasis (see Maynard–Smith 1983). However, one may ask how many characters can be controlled by stabilizing selection simultaneously to infer whether stabilizing selection is a plausible explanation of stasis in complex organisms.

Consider a population of ideal unconstrained organisms, each comprising n genetically uncorrelated characters. Let us further assume that all these characters have the same variance and heritability and are all under stabilizing selection of equal intensity. This is the most simple situation where we may ask what`s the influence of phenotypic complexity (i.e. the number of characters) on the stability of the population (`sphere model`).

According to deterministic population genetic theory there will be no influence of phenotype complexity on the stability of the population. After a pertupation the population will converge towards the optimum at a rate which is independent of n. The only effect of increasing complexity would be that mean fitness w decreases. If fitness differences are only due to variability differences, mean fitness has to be above a certain threshold in order to avoid extinction. The threshold value depends on the birth rate. The higher birth rate, the lower the threshold–value of mean fitness. Hence, theoretically, low mean fitness can be compensated by higher birth rates.

In finite populations chance effects will have the tendency to drive the population away from the optimum in phenotype space. The average phenotype of the population will therefore not approach the optimum but will, on the average, be found at a certain distance apart from the optimum. The average distance depends on the intensity of stabilizing selection, the amount of variance and population size (Lande 1976)

If chance effects are taken into account only in reproduction the average distance from optimum will increase with the square root of n. This conclusion is easily obtained from the results of Lande (1976). In fig. 2 this prediction is compared with Monte Carlo simulations. It is evident that the simulations are in good agreement with the predictions as long as the number of characters is low (less than 10). For more complex phenotypes the simulations tend to deviate from the predictions. The average phenotype of the population has a greater distance from optimum than predicted.

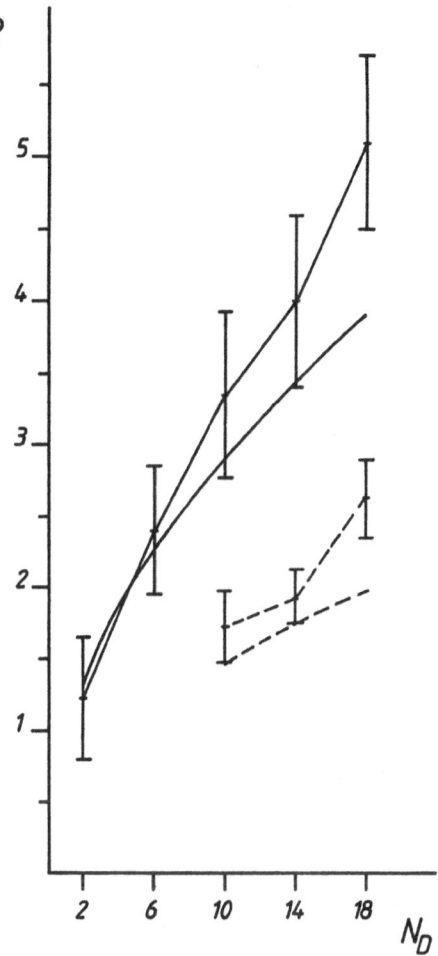

Fig.2: Simulation results of the sphere model. Average distance R of the mean phenotype from optimum is plotted as a function of N_D, the number of characters under stabilizing selection. The solid graph gives the predictions and results for a population size N_e=20, and the dotted graph those for a population with N_e=80. The smooth curve in each graph gives the predictions according to Lande's model. Note that the simulation results tend to deviate from the predictions as the number of characters becomes large. If more than ten characters are under stabilizing selection the average distance is greater than predicted.

The reason of this effect lies in chance effects during selection. In Lande's model the life cycle consists of two stages: selection and reproduction. Neglecting chance effects during selection, leads only to slight errors as long as the number of characters is small. However, it could be shown analytically (Wagner, unpublished) that the variance of the selection step increases exponentially with the number of characters.

This is a consequence of low mean fitness, caused by simultaneous selection at many characters. The variance of selection steps increases also with increasing distance from optimum.

This effect cannot be compensated by increasing the birth rate, since it results from stochastic effects that are independent of offspring number. Hence, complex phenotypes tend to run out of control of stabilizing selection as the number of independently varying characters becomes large. Constraints, i.e. a reduction in the number of independently varying characters, appear to be required for natural selection to be effective. Only constrained phenotypes can be effectively controlled by stabilizing selection.

CONCLUSIONS

In section 2 of this paper empirical evidence is reviewed suggesting that organismic evolution is an interplay between constraints and adaptive opportunities. This fact is no direct consequence of basic population genetic laws. Only if population genetic mechanisms are considered in a systems context, the significance of constraints can be evaluated. Constraints on phenotypic variation can be a prerequisite of adaptive evolution if the characters are functionally coupled. Furthermore the effectivity of stabilizing selection on a character is also a systems property, depending on the variation of all characters that are under stabilizing selection simultaneously. These results, though not sufficient to be called "systems theory of evolution", should exemplify how a systems analytical approach can complement population genetic theory. Systems theory of evolution tries to explain, in population genetic terms, what the difference is between the evolution of an organism and the population genetics of a single character or gene.

Bibliography

Alberch, P. (1980) Ontogenesis and morphological diversification. Amer. Zool. 20, 653–667.

Alberch, P., Gale, E. A. (1983) Size dependency during the development of the amphibian foot. Colchicine-induced digital law and reduction. J. Embryol. exp. Morph. 76, 177–197.

Arthur, W. (1984) Mechanisms of morphological evolution. J. Wiley, Chichester.

Bakker, R. T. (1983) The deer flees, the wolf persues: Incongruencies in predator-prey coevolution. In: D. J. Futuyama and M. Slatkin (Eds.) Coevolution, Sinauer Ass Inc., Sunderland Mass, pp 350–382.

Bookstein, F.L., Gingerich, P. D., Kluge, A. G. (1978) Hierarchical linear modeling of the tempo and mode of evolution. Paleobiology 4, 130–134.

Bürger, R. (1985) Constraints for the evolution of functionally coupled characters. Evolution, 40, 182–193.

Charlesworth, B., Lande, R., Slatkin, M. (1982) A neodarwinian commentary on macroevolution. Evolution 36, 474–498.

Cherry, L.M., Case, S. M., Kunkel, G. G., Wyles, J. S., Wilson, A. C. (1982) Body shape metrics and organismal evolution. Evolution , 36, 914–933.

Cheverud, J. M. (1982) Phenotypic, genetic, and environmental morphological integration in the cranium. Evolution 36, 499–516.

Croin, J. E., Boaz, N. T., Stringer, C. B., Rak, Y. (1981) Tempo and mode in hominid evolution. Natrue 292, 113–122.

Douglas, M. E., Avise, J. C. (1982) Speciation rates and morphological divergence in fishes: Tests of gradual versus rectangular modes of evolutionary change. Evolution 36. 224–232.

Eldredge, N. (1971) The allopatric model and phylogeny in Paleozoic invertebrates. Evolution 25, 156–167.

Eldredge, N., Gould, S. J. (1972) Punctuated equilibria: An alternative to phyletic gradualism. In: T. J. M. Schopf, Models in Paleobiology, pp 82–115.

Gingerich, P. D. (1974) Stratigraphic record of early Eocene *Hyopsodus* and the Geometry of mammalian phylogeny. Nature 248, 107–109.

Haken, H. (1977) Synergetics. Springer-Verlag, Berlin, Heidelberg and New York.

Johanson, D. C., White, T, D. (1979) A systematic assessment of early african Hominids. Science **203**, 321-330.

Kaufmann, R. (1933) Variationsstatistische Untersuchung über die "Artabwandlung" und "Artumbildung" an der Oberkambrischen Trilobitengattung *Olenus* DALM. Abh. Geol. Palae. Inst. Univ. Greifswald 10, 1-55.

Kauffmann. S. A. (1983) Developmental constraints: internal factors in evolution. In: B. C. Goodwin, N. Holder, C. C. Wylie (eds.) Development and Evolution. Cambridge University Press, Cambridge. pp. 195-225.

Kurtén, B. (1959) Rates of evolution in fossil mammals. Cold Spring Harbor Symp. Quant. Biol. 24, 205-214.

Lande, R. (1976) Natural selection and random drift in phenotypic evolution. Evolution 30, 314-334.

Maglio, V. J. (1973) Origin and evolution of the Elephantidae. Trans. Am. Phil. Soc. NS 63 (3), 1-149.

Malmgren. B. A., Berggren. W. A., Lohmann, G. P. (1984) Species formation through punctuated gradualism in planctonic foraminifera. Science **225**, 317-319.

Maynard Smith. J. (1978) The evolution of sex. Cambridge Univ. Press. Cambridge.

Maynard Smith, J. (1983) The genetics of stasis and punctuation. Ann. Rev. Genet. 17, 11-25.

Mayr. E. (1982) The growth of biological thought. Harvard Univ. Press, Cambridge, MA.

Olson, E. C., Miller, R. L. (1958) Morphological Integration. Univ. of Chicago Press, Chicago.

Ovcharenko, V. N. (1969) Transitional forms and species differentiation of brachiopodes. Paleontol. J. 1, 67-73.

Popper, K. R. (1979) Das Elend des Historizismus. 5. Aufl., J.C.B. Mohr-Siebeck, Tübingen.

Radinsky, L. (1984) Ontogeny and phylogeny in horse skull evolution. Evolution 38, 1-5.

Raup, D. M., Crick, R. E. (1981) Evolution of single characters in the jurassic ammonite **Kosmoceras**. Paelobiology 7, 200–215.

Rechenberg, I. (1973) Evolutionsstrategie: Optimierung technischer Systeme nach Prinzipien der biologischen Evolution. Stuttgart–Bad Cannstatt, Friedrich Frommann Verl.

Riedl, R. (1979) Order in living organisms. J. Wiley, New York.

Rightmire, G. P. (1981) Patterns in the evolution of **Homo erectus**. Paleobiology 7, 241–246.

Rose, K. D., Bown, T. M. (1984) Gradual phyletic evolution at the generic level in early Eocene omomyid primates. Nature 309,.250–252.

Sander, K. (1983) The evolution of pattering mechanisms: Geanings from insect embryogenesis. In: B. C. Goodwin, N. Holder, C. C. Wylie (eds.) Development and Evolution. Cambridge University Press, Cambridge. pp. 137–159.

Seilacher, A. (1973) Fabricational noise in adaptive morphology. Systematic Zoology 22, 451–465.

Schopf, T. J. M. (1982) A critical assessment of punctuated equilibria I. Duration of taxa. Evolution 366, 1144–1157.

Simpson, G. G. (1953) The Major Features of Evolution. Columbia Univ. Press, New York.

Starck, D. (1975) Embryologie. Thieme Verl., Stuttgart und New–York.

Stearns, S. C. (1982) The role of development in the evolution of life histories. In: J. T. Bonner (Ed.) Evolution and development, Springer Verl. Berlin, Heidelberg, New York, pp 237–258.

Stearns, S. C. (1983) The influence of size and phylogeny on patterns of covariation among life-history traits in the mammals. OIKOS 41, 173–187.

Stenseth, N. C., Maynard Smith, J. (1984) Coevolution in ecosystems: Red Queen Evolution or stasis. Evolution 38, 870–880.

Van Valen, L. (1973) A new evolutionary law. Evol. Theory 1, 1–30.

Vermeij, G. (1973) Biological versatility and earth history. Proc. Nat. Acad. Sci. USA 70, 1936–1938.

Wagner, G. P. (1982) The logical structure of irreversible systems transformations: A theorem concerning Dollo's Law and chaotic movement. J. theor. Biol. **96**, 337–346.

Wagner, G. P. (1983) On the necessity of a systems theory of evolution and its population genetic foundation: Comments on Dr. Regelmann's article. Acta Biotheoretica **32**, 223–226.

Wagner, G. P. (1984a) Coevolution of functionally constrained characters: Prerequisites of adaptive versatility. BioSystems **17**, 51–55.

Wagner, G. P. (1984b) On the eigenvalue distribution of genetic and phenotypic dispersion matrices: Evidence for a nonrandom organization of quantitative character variation. J. Math. Biol., **21**, 77–95.

Wagner, G. P. (1985) Über die populationsgenetischen Grundlagen einer Systemtheorie der Evolution. In: J. A. Ott, G. P. Wagner, F. M. Wuketits (Hg.) Evolution, Ordnung und Erkenntnis. Verlag P. Parey, Berlin und Hamburg, pp. 97–111.

Wagner, G. P. (1986). The systems approach: An interface between developmental and population genetic aspects of Evolution. In: D. M. Raup and D. Jablosnki Patterns and Processes in the History of Life, Springer. Berlin-Heidelberg.pp.149–165.

Wagner, G. P. (1988a) The influence of variation and developmental constraints on the rate of multivariate phenotypic evolution. J. Evol. Biol. **1**, 45–66.

Wagner, G. P. (1988b) The significance of developmental constraints for phenotypical Evolution by natural selection. In: G. de Jong (Ed.) Population Genetics and Evolution. Springer Verl., pp. 222–229.

Wagner, G. P. (1988c) The Gene and its Phenotype. Biol. & Phil. **3**, 105–115.

Wake, D. B. (1981) The application of allozyme evidence to problems in the evolution of morphology. In: G. G. E. Scudder and J. L. Reveal (eds.) Evolution today, Proceedings of the second international congress of systematic and evolutionary biology, pp. 257–270.

Wake, D. B., Roth, G., Wake, M. H. (1983) On the problem of stasis in organismal evolution. J. theor. Biol. **101**, 211–224.

Washburn, S. L. (1960) Tools and Human Evolution. Scientific American **203** (3), 63–75.

Williamson, P. G. (1981) Palaeontological documentation of speciation in Cenozoic molluscs from Turkana basin. Nature 293, 437–443.

Wolpert, L. (1982) Pattern formation and change. In: J. T. Bonner (ed.) Evolution and Development. Springer Verl., Berlin, Heidelberg, New York. pp. 169–188.

Wuketits, S. (1983) Biologische Erkenntnis: Grundlagen und Probleme. Gustav Fischer, Stuttgart.

Index of Subjects

THEORY AND DECISION LIBRARY

SERIES A: PHILOSOPHY AND METHODOLOGY OF THE SOCIAL
SCIENCES

Already published:

Conscience: An Interdisciplinary View
Edited by Gerhard Zecha and Paul Weingartner
ISBN 90–277–2452–0

Cognitive Strategies in Stochastic Thinking
by Roland W. Scholz
ISBN 90–277–2454–7

Comparing Voting Systems
by Hannu Nurmi
ISBN 90–277–2600–0

Evolutionary Theory in Social Science
Edited by Michael Schmid and Franz M. Wuketits
ISBN 90–277–2612–4

The Metaphysics of Liberty
by Frank Forman
ISBN 0–7923–0080–7

Towards a Strategic Management and Decision Technology
by John W. Sutherland
ISBN 0–7923–0245–1

Social Decision Methodology for Technological Projects
Edited by Charles Vlek and George Cvetkovich
ISBN 0–7923–0371–7